ENVIRONMENTAL HAZARDS

RADIOACTIVE MATERIALS AND WASTES

ENVIRONMENTAL HAZARDS

RADIOACTIVE MATERIALS AND WASTES

A Reference Handbook

E. Willard Miller
Department of Geography

Ruby M. Miller
Pattee Library

The Pennsylvania State University

CONTEMPORARY WORLD ISSUES

ABC-CLIO

Santa Barbara, California
Oxford, England

Library of Congress Cataloging-in-Publication Data

Miller, E. Willard (Eugene Willard), 1915–
Environmental hazards : radioactive materials and wastes : a reference handbook / E. Willard Miller, Ruby M. Miller.
p. cm. — (Contemporary world issues)
Includes bibliographical references.
1. Radioactive pollution. 2. Radiation. I. Miller, Ruby M. II. Title. III. Series.
TD196.R3M55 1990 363.17'99—dc20 90-34159

ISBN 0-87436-234-2 (alk. paper)

97 96 95 94 93 92 91 10 9 8 7 6 5 4 3 2

ABC-CLIO, Inc.
130 Cremona Drive, P.O. Box 1911
Santa Barbara, California 93116-1911

Clio Press Ltd.
55 St. Thomas' Street
Oxford, OX1 1JG, England

This book is Smyth-sewn and printed on acid-free paper ∞.
Manufactured in the United States of America

Contents

Preface

THE STUDY OF ATOMIC PHYSICS and chemistry began when the German chemist Martin Klaproth isolated uranium in 1789. In the late nineteenth century, medical uses of radioactive materials were developed. The invention of the X-ray machine and the use of radioactivity to treat cancer were major contributions to medical science.

In the twentieth century, scientists recognized that the atom possessed a tremendous amount of energy. The energy within the atom was dramatically demonstrated by the explosion of the first atomic bomb in the early 1940s. Soon after this event, it was recognized that the release of energy from the atom could be controlled by a nuclear reactor, making it possible to produce electricity.

With the explosion of the first atomic bomb, the environmental hazards from radioactivity became immediately evident. In the production of electricity from nuclear materials, a desirable product (electricity) results, but the process may also cause radioactive contamination of the environment, resulting in great health risks.

Since the 1940s, a wide spectrum of individuals has become concerned about the use of the atom for both military and civilian purposes. Many scientists and engineers have questioned the use of the atom in nuclear warfare and for producing electricity. Social scientists, from local to international levels, are also concerned with the military use of the atom as well as with nuclear energy developments. But, possibly most important, individuals throughout the world have become aware of the potential dangers of radiation from nuclear activities and question the use of the power of the atom in the future. Nevertheless, the use of the atom to produce atomic weapons and nuclear energy remains unresolved. This may be the most important question facing mankind today.

This volume begins with a description of the nature of radiation. A discussion of natural sources of radiation follows, with an emphasis on the health hazards of radon. The man-made sources—atomic explosions and nuclear energy—are then treated in detail. The completely destructive effects of atomic explosions are explored. In the production of nuclear energy, the dangers of radiation are investigated in the nuclear fuel cycle from the mining of uranium to the final disposal of spent radioactive materials. The first chapter concludes with a discussion of reactor plant accidents and the health effects of radiation exposure.

As a response to the dangers from radiation that became evident after the creation of the atomic bomb and the development of nuclear energy, the legislation that has evolved to protect the general public at national and international levels is the subject of Chapter 3. The initial Atomic Energy Act of 1946 in the United States established a national policy for the development and regulation of the atomic industry for both nuclear weapons and peaceful operations. More recent legislation has concentrated on transportation of radioactive materials and the disposal of wastes and their effects on human health and safety. From the time of the explosion of the first atomic bomb, it was recognized that international control of nuclear weapons was a necessity for a peaceful world. This chapter concludes with a discussion of international regulation and the attempts at creation of nonproliferation zones.

Other chapters provide information on a variety of topics. The chronology in Chapter 2 lists some of the critical dates, from the discovery of uranium to the modern utilization of the energy from the atom, and the environmental hazards that result from these endeavors. Many organizations have been established to consider the use of atomic energy, and Chapter 4 provides an organizational directory, divided into three major parts. The first considers the private organizations, the second lists U.S. government agencies of which the Nuclear Regulatory Commission is the major one, and the third lists international organizations.

In recent years there has been a massive increase in the amount of literature on the use of radioactive materials. The literature varies from scientific to politically oriented studies. Chapter 5's selected bibliography, including about 100 annotated book citations and several hundred journal articles and government documents, is organized under three basic categories—radioactivity, nuclear warfare, and nuclear energy. At the end of the bibliography, there is a list of selected journals that publish articles on nuclear activities. The volume concludes with an annotated list of films dealing with radioactivity and a short glossary.

E. Willard Miller
Ruby M. Miller
The Pennsylvania State University

1

Radioactive Waste: A Perspective

THE USE OF RADIOACTIVE MATERIAL in today's highly technical society has not only a positive but also a negative connotation. On the positive side, radiation has been used in medical practice to diagnose and treat certain diseases, such as cancer, for nearly a century. In the twentieth century, when the splitting of the atom became a reality, two new uses developed. When the energy was released in a single burst, an atomic bomb resulted, the most devastating weapon the world has ever conceived. When the energy of the atom was released slowly in a nuclear reactor, the energy was converted into electrical power.

In each of these uses, however, radioactive wastes are produced that have a life span of a few minutes to tens of thousands of years. When an atomic bomb is detonated in the atmosphere, an unbelievable amount of radioactive waste is released. Because of the grave adverse consequences to the health of vast numbers of the world's peoples, most nations have now banned atomic bomb testing in the atmosphere. Nevertheless, the proliferation of the ability to build an atomic bomb provides the potential for the greatest environmental hazard facing mankind today. Dire predictions indicate that the radioactive wastes created by a nuclear war could destroy civilization as it is known today.

In the development of atomic energy two dangers exist from radioactive wastes. First, if an accident occurs in an atomic reactor, such as at Chernobyl in 1986, radioactive wastes can contaminate the atmosphere and land areas hundreds of miles from the site of the nuclear plant accident. Second, in the production of energy using uranium as a fuel, radioactive wastes occur in all stages of the fuel cycle, from mining to the production of electrical energy. Most critical in this process is the radioactive waste remaining in the spent fuel cell. The ultimate disposal of such waste is one of the critical issues facing the future development of nuclear energy.

In the medical use of radioactive materials there are also radioactive wastes produced. Most of this waste is low level and has a short life span, but its distribution is widespread and huge quantities are produced. Although the U.S. Congress has passed legislation indicating the disposal of this waste must be dealt with at the state level, no final solution has been reached. Because the waste is radioactive, the attitude of communities to local disposal sites has been "not in my backyard."

The presence of radioactive wastes in the environment is a twentieth-century phenomenon. Because of the wide potential danger to human health, political activity to limit the use of radioactive material has developed. Antinuclear and antiproliferation groups have evolved to stop the development of atomic energy and the proliferation of the atomic bomb.

Radiation

There is evidence in the environment that radiation dates from the beginning of time. Life has thus developed with an ever-present background of radiation. From a philosophical viewpoint, as early as 20 B.C. the Greek philosopher Democritus taught that all matter consisted of tiny particles. It was not, however, until late in the nineteenth century that scientific investigation began to identify and understand radiation.

In recent times few subjects have evoked such public attention as ionizing radiation and its risks to individuals. It must be understood that radiation is everywhere. It is a part of our lives from the moment of our existence to the grave. Only under certain exposure conditions is it harmful to human health.

Nature of Radiation

Radiation describes the process in which energy, in the form of rays of light or heat, is emitted from atoms and molecules as they undergo change. Radiation may originate from natural sources or from man-made sources. When considering radiation there are four energy sources—alpha (α), beta (β), gamma (γ), neutrons (η). The energy for each of these varies greatly.

The penetrating abilities of nuclear radiation vary greatly. Alpha particles travel only a few inches and can be stopped by skin tissue or by a heavy sheet of paper but are very dangerous when substances emitting them are ingested or inhaled. Beta move a few feet through air and can be stopped by a quarter inch of wood or a three-tenths inch sheet of aluminum. Like alpha particles, they can be harmful when emitted inside the body. In contrast, gamma rays travel hundreds of feet through the atmosphere and can penetrate a concrete block or several inches of lead. These gamma rays of electromagnetic energy can damage tissue. Neutrons also travel hundreds of feet in the air and can be stopped by specially enforced concrete and water. When neutrons collide with an atom, they can induce radioactivity by combining with its nucleus. The penetration power of radiation is of extreme importance to humans because of the biological effects each emission can cause.

Natural Sources of Radiation

Natural sources provide by far the greatest proportion of radiation received by individuals. Exposure to radiation is inescapable. Radiation comes to the earth from outer space and rises from radioactive material in the earth's crust. An individual is subject to irradiation that remains outside the body as well as radiation that is inhaled in air or swallowed in food or water. Although everyone on the planet is subject to natural radiation, some people receive much more than others. If an individual lives in an area of radioactive rocks or soils, the dose of radiation is larger. The dose may also depend on life-style. If an individual spends a large amount of time on a beach exposed to the sun, the radiation dose increases.

Cosmic Rays

About one-half of the radiation received by an individual comes from exposure to cosmic rays. These rays originate from the sun and from interstellar space. The rays consist of high-energy particles of which about 70 percent are protons, 20 percent alpha particles, and 10 percent a variety of nuclei. The poles receive more cosmic rays than the equatorial regions because the earth's magnetic field directs the radiation toward the polar regions.

Secondary cosmic rays are produced by interactions of the primary rays and atmospheric nuclei and consist largely of mesons, electrons, protons, neutrons, and gamma rays. At sea level, nearly all cosmic radiation is from secondary sources, with 80 percent being mesons and 20 percent electrons. (See the Glossary for explanations of terms.)

The intensity of cosmic radiation also increases with altitude. A person living at sea level will receive an effective dose equivalent to about 300 microsieverts of cosmic radiation each year, but at 6,000 feet the dose will be several times that amount. A passenger on a jet airplane flying at 30,000 feet from New York to Paris will be exposed to about 50 microsieverts on a normal flight.

Cosmogenic Radionuclides

Approximately 20 radionuclides are known to be continuously produced in the atmosphere by cosmic ray interaction with air. The natural production of cosmogenic radionuclides in the atmosphere exhibits altitudinal and latitudinal patterns similar to those of cosmic ray intensities. About 70 percent of cosmogenic radionuclides arise in the stratosphere, whereas about 30 percent originate in the troposphere.

With the exception of tritium and carbon 14, the cosmogenic radionuclides contribute very little to the environmental radiation levels. Both carbon 14 and tritium are weak beta emitters and are important only when they become part of the body through our food, water, and air. They then enter body tissues. The estimated annual dose from these two radionuclides is about 10 sieverts per year, with the major portion coming from carbon 14.

Primordial Radionuclides

Primordial radionuclides are those that came into existence at the time of the formation of the earth. These radionuclides can be divided into two groups: those whose half-lives are larger than the age of the earth and those that are radioactive products of a few of these long-lived radionuclides. The first group has about 20 known radionuclides, with half-lives ranging from 7×10^8 years for uranium 235 to about 10 years for dysprosium 156. The second group contains 40 radionuclides, all of which are decay products of uranium 235, uranium 238, or thorium 232. Each of these three radionuclides undergoes successive alpha and beta decay steps until it reaches one of the stable isotopes of lead.

Of the first group of radionuclides, potassium 40 is most important. It is a major source of radiation. Potassium is one of the major elements in the earth's crust, and 0.012 percent of all potassium atoms are potassium 40. This radionuclide is more abundant in the earth's crust than copper, lead, zinc, or cobalt. As a gamma ray emitter, this nuclide provides a significant portion of the terrestrial irradiation. Because potassium is an essential element in the human body, this radionuclide contributes more than half the average internal irradiation dose to the body through release of gamma and beta rays.

The widespread distribution of uranium and thorium throughout the earth's crust leads to a significant contribution to the external irradiation dose. Because uranium 238 is unstable, the particles in its nucleus are altered as protons and neutrons break away. Uranium 238 goes through a series of changes, becoming first thorium 234, then protactinium 234, and on to become radon 222 and ultimately lead 206.

Radon

Of the natural sources of radiation, radon may be the most important. It is a tasteless, odorless, invisible gas, seven and one-half times heavier than air. The United Nations Scientific Committee on the Effects of Atomic Radiation estimates that radon, together with its daughters—radionuclides formed as it decays—contributes about three-quarters of the annual effective dose equivalent received by individuals from all natural sources. Most of the dose results from breathing in the radionuclides, particularly indoors.

The two major radon daughters are radon 222, a radionuclide formed by the decay of uranium 238, and radon 220, a product of the decay of thorium 232. Radon 222 is about 20 times more important than radon 220. The radon 222 isotope has a half-life of 3.8 days.

The radon content of regions varies considerably throughout the world depending upon the uranium level in the bedrock and soil. The U.S. Environmental Protection Agency (EPA) has estimated that the average soil contains about 1 part per million of uranium. Phosphate rock contains 50 to 125 parts per million, and granite contains as much as 50 parts per million. In areas of friable soils encouraging high gas mobility, it is generally recommended that radon resistant housing measures be taken.

Outdoor air concentrations of radon are so small that the possibility of their being a health hazard is extremely small, but indoor concentrations are typically two to more than ten times higher. Radon in houses arises primarily from radon that migrates from the soil under the house or in some cases from materials that make up the house, such as cement blocks with a radon concentration. The distribution of indoor concentrations across the United States is widespread, varying from one-tenth of typical levels to a factor of 100 times higher. The indoor concentrations are controlled by balancing between the rate of entry from sources and the rate of removal by ventilation.

It has been determined that the entrance of radon into a building depends upon the radon production rate in the soil, meteorological factors, soil permeability, the type of building substructures, and the stack effect. The "stack effect" is the result of lower pressure inside a building, which causes radon to move from the earth to the lower pressure. The stack effect is a major contribution to air infiltration and can influence the release of radon in soil adjacent to a building and thus cause higher radon entry rates. Wind may also remove radon from the soil, not because of the new inflow of soil gas, but because it causes a greater exchange of air between the building and the soil.

Evidence indicates that a potentially large number of houses, particularly those recently built and those reinsulated to conserve energy, have high indoor radon levels. Measurements in homes where concentrations are an order of magnitude greater than the average suggest an individual risk of lung cancer exceeding 1 percent. Radon levels in houses can be reduced by sealing floors and walls. Covering walls with plastic materials like polyamide, polyvinylchloride, polyethylene, and epoxy paint—or giving several coats of oil-based paints—reduces the emission of radon by as much as tenfold. Even wallpaper may reduce it by 30 percent or more. In addition, using fans to ventilate crawl spaces is a particularly effective way of reducing the radon that does seep into a building.

Water and natural gas are less important sources of radon than the soil, but cannot be ignored. Usually the amounts in these sources is small, but some, especially deep wells, have high concentrations. Radioactive water supplies have radon activity varying from nearly nothing to 100 million becquerels per cubic meter. The greatest danger from water is in hot drinks like tea and coffee. Boiling water before drinking releases most of the radon. In a French study it was found that radon concentrations in bathrooms were about three times higher than in kitchens.

Radon is absorbed by natural gas in the ground. Processing and storage removes much of the radon before it reaches the customer, but concentrations can be high and increase significantly if the gas is burned in unvented stoves. If the gas is processed, much of the radon ends up in liquefied petroleum gas (LPG), a by-product.

Energy conservation procedures can greatly increase radon concentrations. Insulating houses and reducing ventilation reduces the escape of radon daughters. Although this conserves heat, it encourages radon buildup. The increasing use of insulation in Sweden has resulted in a rise of radon in the home. In 1966 a survey revealed no problems of radon in Sweden. Between 1950 and 1970 ventilation rates were cut by half and radon concentrations tripled.

Recent surveys in a number of countries revealed that radon daughter concentration levels ranged from 1,000 to 10,000 becquerels per cubic meter, or from 0.01 to 0.1 percent. This means that many people may be exposed to high concentrations of radon in their homes.

Health Risks from Radon

The principal health risk comes from the inhalation of the short-lived radon 222, and the second greatest risk from breathing radon. The radon daughters are most dangerous because of their short life, because they decay at the place of original deposition in the respiratory system. The membranes of the lungs are particularly sensitive to the carcinogenic effects of radiation. Other environmental factors such as exposure to mineral dust and chemical aerosols may increase the damage of surface cells, thus increasing the sensitivity to radiation.

Because radon and radon daughters can cause cancer, they are a major health concern. Acute exposure can damage organisms and affect the gastrointestinal and nerve systems, resulting in internal bleeding, cardiac distress, and possibly death. Exposure to low doses of radon may produce delayed effects as cells are transformed. Mutations can be deleterious in many ways, such as by reducing resistance to diseases, including cancer.

A study at the University of California at Berkeley showed that radon-exposed individuals who smoked had a much higher cancer rate than nonsmokers. It was estimated that of 1,000 male nonsmokers exposed to radon, 16 would die of lung cancer, 5 more than in a nonexposed group. Among 1,000 female nonsmokers, radon exposure increased the rate from 6 to 9. The statistics are much grimmer for smokers. Of 1,000 males exposed to excess radon, 172 will die of lung cancer, 49 more than among nonexposed male smokers. For women smokers, radon raises the number from 60 to about 85. Why radon is more lethal for men than for women is unknown, and why smokers are at greater risk is also unknown. Some experts believe that smoker-damaged lungs trap the radiation radon particles more effectively.

Other Sources of Natural Radiation

Coal is a natural source of primordial radionuclides. When coal is burned, traces of radioactivity are released. Although the concentration can vary 100-fold from seam to seam, the amount of radioactive material in coal averages less than that found in the earth's crust. When coal burns, however, this radioactivity is largely concentrated in the ash.

Some of the ash remains in the furnace, but a large proportion of the lighter fly ash is carried up the chimney. The cloud from the chimney gradually drifts to the ground, where it irradiates people and settles on the soil. It is estimated that each year the world's coal power stations produce a collective effective dose equivalent of about 2,000 man-sieverts. In addition, stoves and fireplaces may well emit more ash to the atmosphere than power stations. Estimates indicate that these sources produce a collective effective dose equivalent of 100,000 man-sieverts throughout the world. Fly ash is also used to produce concrete blocks, in road building, and to improve soil fertility. Each of these adds radiation to the area.

Phosphate rock is another source of natural radioactive materials. Most of the deposits being exploited contain high concentrations of uranium. In the mining and processing of the ore, radon is released. When the phosphate is applied as fertilizer, the radiation contamination is slight, but it becomes greater if the fertilizer is liquefied or if the phosphate is fed to animals. For example, dairy cattle that graze in pastures fertilized by phosphate can produce milk with significantly increased levels of radioactivity.

Geothermal energy is another source of increased radiation. This source of energy is in an initial stage of development. Heat originates in bedrock, and small quantities of radioactive material are incorporated in the steam. A study in Italy found that emissions from three hot-water geothermal springs produced a collective dose equivalent of 6 man-sieverts per gigawatt year of electricity output—three times the corresponding dose produced by coal-fired power stations. Although geothermal energy makes only the tiniest contribution to the world's energy today, it will become an important source in the future.

Naturally, the levels of primordial radiation vary greatly over the world. In general, the amount of radiation has essentially no effect on the health of the individual in most places. Studies in the United States, France, West Germany, Italy, and Japan indicate that 95 percent of the people live in areas where the average annual dose rate varies from only 0.3 to 0.6 millisieverts. There are a number of spots on the earth, however, where the concentrations of radionuclides are much higher, and health for certain individuals may be endangered. For example, in the city of Pocos de Caldas near São Paulo in Brazil, researchers discovered radiation dose rates at about 800 times the average—250 millisieverts a year. On the southwest coast of India in a 55-mile strip in a thorium-rich sand area, radiation is high. In a study of 8,513 people, the average exposure was 3.8 millisieverts per person per year, but about 60 persons absorbed more than 17 millisieverts.

Man-Made Sources of Radiation

Since the discovery of man-made sources of radiation by the Curies in the late nineteenth century, these sources have greatly increased. For the individual, medical uses of radiation provide the greatest danger. Radiation is used for both diagnosing and treating disease. Since the 1940s and the development of the atomic bomb, and later the use of nuclear fuel to produce energy, there has been the potential for increased radiation in the environment.

Atomic Explosion

Origin of the Atomic Bomb

The first significant event leading to the explosion of an atomic bomb on July 16, 1945, occurred in 1789 when Martin Heinrich Klaproth discovered the element uranium during the examination of pitchblende. From this event scientists gradually developed an understanding of the structure of the atom. A quantum leap forward came in 1905 when Albert Einstein conceived his revolutionary theory of motion. This theory concludes that the mass of an object increases with its speed and is stated by the now famous formula $E = MC^2$, which expresses the equivalence of mass and energy. Without this understanding, nuclear energy could not be achieved.

By 1932 it was well known that the radiation emitted from certain minerals and elements originated from the various structures of the atom. It was then theorized that if the atom could be "split" there would be the release of energy in a chain reaction. By 1938 advances in the laboratory had proved the atom could be split. The stage was now set for the construction of the atomic bomb. The outbreak of World War II triggered the project to build an atom bomb. In 1942 J. Robert Oppenheimer, a University of California physicist, assembled a group of scientists to produce an atom bomb. The project was known officially as the Manhattan Project.

Although the scientists were reasonably well acquainted with the physics involved, putting their theories into practice was another thing entirely. The slightest miscalculations would create an enormous safety hazard for the scientists. For example, neutrons randomly emitted during normal decay of a larger mass of uranium would encounter a larger number of atomic nuclei before they reached the surface than would neutrons emitted from smaller amounts of uranium. The greater the number of nuclei, the greater the number of neutrons emitted to strike even more nuclei. This could result in a lethal dose of radiation. The scientists were fond of saying, "First it goes critical, and then you go critical." No one knew, however, what the critical amount was. Only the most careful trial-and-error experiments gradually solved the problem.

There was an even more critical question scientists were forming. Some scientists were wondering whether the bomb could activate previously undiscovered natural phenomena. For example, some wondered if the bomb might ignite the atmosphere, causing a worldwide conflagration. In an early calculation Edward Teller indicated that in a fusion explosion the heat buildup would be sufficient to set fire to the atmosphere's nitrogen. It was found, however, that Teller had forgotten to account for the heat absorbed by radiation. The chances of the world catching fire were only three in a million. The work continued.

After approximately three years of development an atomic bomb was ready for testing on July 16, 1945. On a tower 100 feet above the ground was a bomb platform known as the Gadget. The site was near Sierra Oscura in the Jarmada del Muerto of New Mexico. The Gadget held 5,000 pounds of explosives surrounding plutonium and wired to precision detonation capable of responding to the slightest variances in electrical charge, all connected to the X-unit, the firing mechanism. At 5:29:45 A.M. the first atomic bomb was detonated.

As the electrical charge raced through the capacitor crushing first the uranium "hammer" and, inside it, the plutonium sphere, the nuclear material became supercritical. At that moment the polonium and beryllium mixed, sending forth a dense spray of neutrons that fissioned nuclei and released still more neutrons. Within a hundred-millionth of a second, a dense neutron cloud formed, fissioning billions of plutonium atoms. At the same time a glow flashed around the top of the tower, the result of radiation from gamma rays and X-rays roaring through the desert air, briefly ionizing it before heating it to millions of degrees centigrade. Within a millionth of a second, 99 percent of the neutrons had passed through the bomb debris and into the atmosphere. The plutonium had already begun to disintegrate to 360 different isotopes and 36 different elements. For a moment, radiation from ultraviolet rays, gamma rays, and X-rays inside the device exerted a pressure of several thousand tons per square inch.

The bomb was now a miniature star eleven times hotter than the sun. The surrounding air was well above 8,000 degrees centigrade. A dense, opaque shield of glowing vapor appeared around the fiery body. A ball of fire enveloped the tower and a pressure wave of 1,000 million atmospheres pounded the earth. The soil was literally vaporized and carried high into the atmosphere. Within less than a second the fireball was more than 1,000 feet high; within seconds it was half a mile wide. It continued to shoot skyward at a tremendous rate. At an altitude of 4 miles the mushroom was orange and pink, and at 7 miles it was a rocking mass of gray ash composed of radioactive debris.

As Oppenheimer watched the explosion, he remembered a passage from the Hindu Bhagavad Gita, "I am become death, shatterer of worlds."

Hiroshima and Nagasaki

Before the atomic bomb test explosion near Los Alamos, New Mexico, a target committee was planning the dropping of the bomb on Japan. After consultation with several officials, including Chief of Staff George Marshall, the committee selected Hiroshima as a good military target because of its many munitions factories.

Hiroshima had never been bombed. Rumor in Japan indicated that Hiroshima was being saved for a special form of destruction. On August 2, 1945, an atomic bomb was placed in the *Enola Gay,* and at 8:15 A.M. the plane dropped the bomb on Hiroshima. At 1,900 feet, the detonation of the bomb formed the initial burning wave with a temperature of over 300,000 degrees centigrade. When the blast hit the ground, a fiery pressure wave moved, creating an oven and incinerating everything in sight. Buildings were vaporized and shattered beyond recognition. Within an hour a black rain of ash began to fall. Thousands had died immediately when the bomb exploded. Many who escaped the trauma of the bomb would die of radiation-induced hemorrhages. The bomb eventually killed 130,000.

On August 8, 1940, a second bomb was dropped on Nagasaki with the same type of devastation. With two cities destroyed, a major confrontation of the forces desiring peace and those wishing to continue the war took place in Tokyo. The debate continued for six days. In a break with tradition that had roots in the seventh century, the emperor addressed the cabinet. The emperors of Japan were considered descendants of the sun-goddess Amaterasu. Normally the emperor's concerns were not of this world. He now was asked for his views on the war. His response presented the viewpoint that the war efforts had been unsuccessful and that, in order to prevent further hardships for the people of Japan, the war should be ended.

Negotiations developed rapidly, and the U.S. war with Japan ended.

Atomic Bomb Testing

At the conclusion of the hostilities of World War II, many military people questioned the value of the atomic bomb. Even General Curtis

Le May, who was in charge of dropping the first bomb on Hiroshima, stated, "The atomic bomb had nothing to do with the end of the war at all." In an effort to prove the value of the most devastating of all weapons, tests were planned by the military.

The testing of atomic bombs to determine their destructiveness began in 1946. The sites selected for the tests were the remote and beautiful Bikini and Eniwetok Atolls lying about 2,700 miles southwest of Hawaii. These isolated islands met the conditions required for testing:

1. They were in an area controlled by the United States in a climatic region free from storms and cold temperatures.
2. They had sheltered areas for anchoring target vessels and measuring the effects of radiation.
3. They were areas with small population that could be readily moved to another area.

The Bikinians agreed to sacrifice their island in order to provide information to the armed forces of the United States. They were moved to the island of Rongerik, in the Marshall Islands, in March 1946, and testing began. Between 1946 and 1958 the U.S. Air Force conducted 66 nuclear tests on Bikini and Eniwetok.

In the late 1940s the army and navy also wanted to test atomic weapons. Although Bikini was an ideal testing site, it was expensive. Also, the air force tests on Bikini were of large bombs, while the army and navy wanted smaller atomic weapons. As a consequence, the Atomic Energy Commission (AEC) agreed to land tests in the desert areas of the Southwest, which became known as the Nevada Testing Grounds. At the same time, the Soviet Union and Great Britain began testing atomic weapons in the atmosphere.

In 1952, between April 1 and June 5, a series of land tests called Tumbler-Snapper was conducted at the Nevada Testing Grounds. The Tumbler tests were devised to measure the effectiveness of the new weapons. The bombs were air-dropped over the target while scientists on the ground about a mile away measured the relationship of burst height and destructiveness. The Snapper tests were designed to improve nuclear weapon design. Beginning in 1953, a new series of atomic tests was prepared. The Upshot-Knothole series consisted of eleven atmospheric detonations, followed by airdrops, seven tower shots, and one warhead fired from an atomic cannon.

Until 1955 few scientists had warned of the adverse effects of radioactive fallout. Most of the concern centered on the possibility of a war involving larger and more lethal destructive weapons. In early 1955 after the Turk detonation, known as Bag Shot, the weather bureau reported that a "harmless radioactive cloud" was hovering over the eastern United States and that radioactivity had been reported in Chicago. Drs. Lanier and Puck from the Colorado Medical Center stated their opposition to the Nevada testing because tests revealed that within a few hours after the Nevada detonation radioactivity increased over the state of Colorado.

Since the beginning of nuclear testing, the AEC had tracked nuclear clouds in their journeys across the country. Then in the mid-1950s, studies were made of chemical reactions to the fallout, with particular attention focused on the 300 isotopes produced in atomic bomb detonations. Questions were raised as to how fission particles affected human beings. The controversy swept through the scientific world. Finally, in 1958 President Eisenhower declared a moratorium on the testing of atomic weapons in the atmosphere. The Soviet Union and Great Britain agreed to the moratorium, and since then these countries have confined their testing of atomic weapons to underground locations. Only France has continued to test atomic bombs in the atmosphere, in the southern Pacific Ocean.

Atomic Bomb Contamination of Bikini

During the testing of bombs on Bikini, the U.S. Navy remained optimistic that as soon as testing was over the inhabitants could return. The dangers of radiation became apparent in 1954. Winds carried the radioactivity from an explosion eastward to the island of Rongelap and Utirik. Ninety percent of the Rongelapese people immediately suffered skin lesions and loss of hair. By the 1980s, 19 of the 21 Rongelapese who were under 12 years of age at the time of the test had developed thyroid tumors or other radiation-related illnesses.

In 1969, 11 years after the atomic bomb testing had concluded, the AEC declared that Bikini had virtually no radiation and there was no discernible effect on plant or animal life. The Bikinians were joyful that they could return to their homeland, and resettlement began. The Department of Interior initially built 40 new homes and planted 50,000 new trees on the island. In 1973 the U.S. government indicated construction was nearly complete, and all Bikinians could return permanently by Christmas.

Problems were becoming evident, however. In 1972 an AEC survey of Eniwetok found that radiation levels were extremely high in certain places. In late 1974 the secretary of interior, Rogers C. B. Morton, alarmed at the findings of routine radiological surveys, halted construction on Bikini Island. In March 1975 Defense Secretary James R. Schlesinger requested a thorough survey of the radioactivity on Bikini. Because of bureaucratic wrangling, the surveys were delayed.

Meanwhile, the Bikinians expressed a desire to build houses in the interior away from the lagoon. A routine survey in June 1975 indicated that the interior was too radioactive for housing and that some wells were contaminated with radioactive plutonium. Further, the survey revealed that although coconuts were likely to be safe, breadfruit and pandanus, two basic food staples, contained unacceptably high levels of radiation.

The Bikinians, frustrated and confused, brought suit in federal court in October 1975 to force the United States to stop the resettlement program until a comprehensive radiological survey to the island could be made. The U.S. government agreed to make the survey, and $2.6 million was appropriated in 1977, but again the Defense Department took no action, saying that more money was needed. The squabble lasted three years. In the meantime, tests in 1977 showed that the levels of strontium 90 in well water on Bikini exceeded acceptable U.S. standards. Medical examinations revealed that the people on Bikini had absorbed cancer-causing radioactive elements such as strontium, plutonium, and cesium.

In 1978 a medical team informed the 139 people living on Bikini that they could no longer eat locally grown food, but must import all food and water. In a single year, the population had shown a 75 percent increase in body burdens of radioactive cesium 137. The people of Bikini may have ingested the largest amount of radiation of any known population. It was concluded that all people had to be removed from the island. In August 1978 the 139 people were removed, and no one has been allowed to live there since. The Bikini experience has raised many questions and provided some answers. Certainly there is positive evidence that radioactive environmental contamination caused by atomic testing persists for many years. There is also strong evidence that the American government was less than responsible in its actions to the Bikinians. There remains the problem of where the Bikinians are to secure a permanent home. In the wider perspective, atomic testing on Bikini illustrates on a small scale the dangers a modern atomic holocaust could create.

Nuclear Energy

The use of nuclear raw materials to produce energy began in 1957 when the first reactor at Shippingport, Pennsylvania, began operations. At that time nuclear power was hailed as a cheap, clean, efficient, and safe energy source for the world. Public opinion for nuclear power dwindled over the years, however, and with the Three Mile Island accident of 1979 and the Chernobyl accident of 1986, many doubts have been raised as to the safety of nuclear power for the general public.

Nuclear Fuel Cycle

There are numerous processes in the nuclear fuel cycle. The main steps involved are:

- Mining of the ore
- Milling
- Conversion
- Enrichment
- Fuel fabrication
- Power production (nuclear energy)
- Fuel reprocessing
- Waste management
- Transportation

At each of these steps there is some danger of exposing the public to harmful radiation. The greatest dangers stem from the possibility of a reactor accident during energy production and the problem of nuclear waste disposal after the energy production takes place.

Mining

Radioactivity in Uranium Mines

As early as the sixteenth century it was recognized that miners who worked for several years in pitchblende mines were subject to fatal pulmonary diseases (Arell S. Schurgan and Thomas C. Hollocher, "Radiation Induced Lung Cancer among Uranium Miners," in Union of Concerned Scientists, *The Nuclear Fuel Cycle* [Cambridge, MA: MIT Press, 1975], pp. 9–40). For example, a higher mortality level from pulmonary disease was recognized in miners from the Erz Mountains of Germany and in Joachimsthal in Czechoslovakia. Many of the miners in these districts died in middle age of a pulmonary disease known as Bergkrankheit, or mountain sickness.

An early study of the Schneeberg mines of Germany conducted between 1869 and 1877 found that the 650 miners working during that period had a life expectancy of 20 years after entering the mines. Two German doctors, Harting and Hesse, found that 75 percent of the deaths were due to lung cancer. The death rate was higher for underground miners than for those who worked on the surface. The doctors surmised incorrectly that the high death rate was due to breathing arsenic. Later studies between 1900 and 1940 provided statistical evidence that the high incidence of lung cancer was due to the presence of radioactive substances in the mines.

A study conducted in the Schneeberg region between 1924 and 1939 revealed very high values of radioactivity, ranging from 3.6 to 180 working levels. (A working level [WL] is defined as any combination of short-lived radon daughters in one liter of air that will result in the ultimate emission of 1.3×10^5 MeV of potential alpha energy.) In a 17-year period with 30 WL as the average exposure, the cumulative was more than 6,120 WL months, or more than 12,000 rad. Without doubt, radiation was the cause of lung cancer at the Schneeberg mines in the Erz Mountains. In a comparable study at Joachimsthal the exposure was 3,840 WL months, or about 8,000 rad.

Prior to 1930, uranium ores were mined for radium in Arizona, Utah, Nevada, Colorado, and New Mexico. Beginning in 1946, large-scale uranium mining took place, first for atomic weapons production and later for electricity. Intensive mining continued through 1968, when the existence of adequate stockpiles of uranium led to a reduction of mining activity. During this period more than 6,000 mines were exposed to radiation gases.

Radon measurements were first made in U.S. uranium mines in 1949. The Colorado Department of Health recognized the possible health hazard to miners and with the cooperation of the U.S. Public Health Service began to monitor the mines for radioactivity. In 1951 a survey of radon daughter concentrations began. The surveys were somewhat spotty and largely carried out to collect information rather than to control radioactivity in mining operations. Over the years evidence mounted that radioactivity was high in uranium mining operations and could be harmful to miners' health. In 1957 the U.S. Public Health Service published the results of radon daughter surveys and predicted a significant mortality rate from lung cancer among miners. As a result, in 1961 major radon control programs were initiated in underground mines. Advances in ventilation technology have continually improved the situation. Prior to 1956 it was common for mines to operate at 5–10 WL, with a few operating at 20–200 WL. By 1968 a majority of the mines showed values of 0.5–2 WL, with about 20 percent at less than 0.5 WL and none greater than 5 WL. Although the risk of acquiring lung cancer has not been eliminated for miners, it has been greatly reduced.

A critical health monitoring program was conducted with 3,400 white and 780 nonwhite (mostly American Indian) miners from 1950 to 1968, and less rigorously up to the 1980s. In 1973 the miners showed an excess of 180 respiratory malignancies. It is predicted that ultimately the group will suffer 600 to 1,100 excess lung cancer deaths due to irradiation. Smoking appears to increase substantially the risk of radiogenic lung cancer, particularly in shortening the time before cancer develops. American Indians appear to have a significantly shorter time before cancer appears than do whites.

The large number of uranium miners contracting lung cancer could have been predicted on the basis of past investigations in Europe. Federal authorities, primarily the AEC, failed to prevent contamination by radioactive materials. The health of the miners was largely ignored in the mining process. The means for controlling radon through vigorous ventilation of mines were known and available, as were the instruments for measuring radiation.

Effects of Radon on Miners

In the mining process, natural decay of uranium 238 results in the formation of various radioactive products in ore bodies. Of these, radon 222 is a noble gas that escapes from rock surfaces within the mine. Radon 222 has a half-life of 3.8 days and decays through alpha (α), beta (β), and gamma (γ) emissions through a set of short-lived nuclides, which as a group exhibit a half-life of about one-half hour before reaching lead 210 (see Table 1).

Radon daughters, either as free ions or absorbed in water or dust particles, are inhaled, deposited, and retained in the respiratory system. Consequently, the respiratory system can be exposed to radiation doses that far exceed body doses from internally dissolved radon. When lung tissues are exposed to radon, the major tissue damage is attributed to alpha particles, which have short range and high linear energy transfer.

Lung damage from radon daughters appears normally in the hilar region of the lung. This is an area that is ciliated and covered with a sheet of mucous. The ciliated cells are interspersed with goblet cells that provide the mucous secretion. Because this surface layer of cells develops from a layer of basal cells attached to the basement membrane, the integrity of the epithelium depends on the continued integrity of the basal cells. Thus, the nuclei of these basal cells become the relevant biological target. Estimates of the distance between the biological target, the basal cell nuclei, and the source, alpha particles on the mucous layer, are 36 microns minimum and 63 microns median thickness. The range of alpha particles from radon daughters (5–7 MeV) in tissue is roughly 39–71 microns. Thus, alpha particles entering the epithelium will deliver on average a significant portion of their energies to the basal cell region. With an effective half-life of one-half hour, it is evident that radon daughters will decay mainly on the bronchial epithelium before they can be removed by normal clearance mechanisms.

TABLE 1

Radioactive Decay of Radon 222

NUCLIDE	HALF-LIFE	MAJOR RADIATION
Ra 222	3.8 days	Alpha
Po 218	3 minutes	Alpha
Pb 214	26.8 minutes	Beta, gamma
Bi 214	19.7 minutes	Beta, gamma
Po 214	0.0002 seconds	Alpha
Pb 210	20.4 years	Beta, gamma

Dust particles larger than about five microns are filtered out in the upper respiratory tract and do not reach the deep air passages. It is thus evident that the actual radiation dosages depend upon a number of factors:

1. Concentration of radon daughters in the atmosphere
2. Respiratory volume and exchange rate
3. Distribution of radon daughters between free ions and ions absorbed by dust particles
4. Size of the dust particles
5. Respiratory clearance rate

The first quantitative relationships between the incidence of respiratory cancer and cumulative dosage of radiation were described by Wagoner and his associates in 1965 (Joseph K. Wagoner and others, *New England Journal of Medicine* 273 [1965], 181). Using WLMs (working-level months) as cumulative radiation exposure estimates, they reported that the incidence of lung cancer increases from 3.1 per 10,000 miners per year to 116 per 10,000 miners per year over six exposure categories ranging from less than 120 WLM to more than 3,720 WLM. Although the confidence limits were wide in this study, the linear dose-response curve made the existence of a causal relationship between airborne radiation and respiratory cancer in U.S. uranium miners virtually certain by 1965.

By the mid-1960s, the following conclusions had been reached. The highest incidences of lung cancer were occurring in miners who:

1. Had worked in mines in which the average concentration of radon daughters was usually greater than 10 WL
2. Had a total cumulative exposure exceeding 1,000 WLM
3. Had mined uranium ore for more than 15 years
4. Were moderate to heavy cigarette smokers

It was thus recognized that uranium miners needed protection. On July 1, 1971, the AEC established standards of radiation exposure for uranium miners. They stipulated that no individual was to receive an exposure of more than 2 WLM in any consecutive 3-month period and no more than 4 WLM in any consecutive 12-month period. The conversion of WLM to rems is roughly 2.5 WLM = 15 rems/year.

These standards have been criticized by a number of scientists. Although no set standards can be absolutely safe, there is no evidence at present that a threshold exists for the induction of lung cancer by alpha radiation in humans. If a threshold does exist, it probably lies below the level of 50 rad.

Nuclear Fuel Reserves

Fuel Reserves for Nuclear Fission

The nuclear energy industry is based on uranium. Natural uranium is composed of 99.274 percent U 238, 0.720 percent U 235 and 0.006 percent U 234. This small portion of U 235 provides the fuel for reactors. A reactor has also been developed that uses thorium to produce energy.

It is difficult to determine precisely the reserves of uranium and thorium in the world. Not only is there no uniform classification, but many nations withhold data, including the Soviet Union, the People's Republic of China, and Eastern Europe. In 1984 it was estimated that the Western world contained uranium reserves of 1,498,586 tons of recoverable ore at less than $80 per kilogram and an additional 713,376 tons recoverable at $80 to $130 dollars per kilogram. The Federal Institute of Geosciences and Natural Resources in Hanover, West Germany, has estimated that the socialist countries have 150,000 to 300,000 tons of proven uranium reserves and additional potential resources of 1,115,000 to 1,630,000 tons. In addition to the proven and potential resources, an estimated additional 5,350,000 to 6,300,000 tons of possible and speculative uranium exists on the planet. The earth thus contains a potential of 7,800,000 to 9,400,000 tons of uranium.

Uranium reserves are widely scattered over the world, but major deposits are concentrated in a few countries (see Table 2). The six leading countries, with production costs of $80 per kilogram or less, have nearly 7 percent of the world's recoverable reserves. Australia, at 30 percent, has the largest recoverable reserve, followed by South Africa (12 percent), Nigeria (11 percent), Brazil (10 percent), Canada (10 percent), and the United States (8 percent). The world's oceans contain a nearly inexhaustible supply of uranium. The concentration, however, is very

TABLE 2

Uranium Reserves

LOCATION	RECOVERABLE AT LESS THAN $80 PER KILOGRAM (1984)	RECOVERABLE AT $80–130 PER KILOGRAM (1984)
Australia	463,000	63,000
South Africa	191,000	122,000
Nigeria	170,400	—
Brazil	163,273	—
Canada	155,000	59,000
United States	131,300	266,800
France	55,953	11,109
India	31,025	10,772
Spain	26,700	6,200
Algeria	25,000	0
Mexico	23,600	—
Gabon	18,700	4,700
Argentina	15,385	3,552
Japan	7,700	0
Portugal	6,700	1,500
Italy	4,800	0
Turkey	2,500	2,100
Sweden	2,000	37,000
Zaire	1,800	—
Chile	1,000	—
Germany, Fed. Rep.	850	4,200
Peru	500	—
Greece	400	—
Syrian Arab Rep.	—	80,000
Denmark	—	27,000
Korea, Rep.	—	10,000
Finland	—	3,400
Somalia	—	660
Austria	—	300
Afghanistan	—	75
Albania	—	8
WORLD	1,498,586	713,376

Source: World Energy Conference.

low and processing costs will be high, so extraction is not likely to occur until well into the twenty-first century.

Thorium, like uranium, is found in the earth. Its average abundance in the earth is 12 ppm, which is about three times that of uranium. Because demand for thorium has been low and reactors are not expected to use it as a fuel in the immediate future, little exploration for deposits has occurred. At the present time, the estimated reserves and additional thorium resources total 3,893,500 tons. This figure could greatly increase when new deposits are sought. Thorium reserves are widely scattered. The United States has about two-thirds of North American reserves, and Canada has the other third. The South American reserves are found exclusively in Brazil. The European reserves are found in Denmark (Greenland) and Norway, the Asian reserves are in India, and the African reserves are in Egypt. Among the socialist countries, the Soviet Union has the largest reserves.

Fuel Reserves for Nuclear Fusion

Although the underlying processes have been known for decades, no laboratory has been able to build a fusion reactor. The first fusion bomb was exploded on November 11, 1952, in the United States, but the extreme pressure and temperature required to detonate it were achieved by the explosion of a fission device. Although it is uncertain when a fusion reactor will become commercially feasible, some estimates have been made of the fuels such a reactor would require—lithium, deuterium, and beryllium.

The Max Planck Institute for Plasmaphysik estimated the economic recoverable reserves of lithium to be 1.4×10^6, representing an energy potential of 35 to 123×10^3 quads, depending upon the reactor. The indicated and inferred reserves are 5.2×10^6 tons, which corresponds to 128 to 455×10^3 quads. The total reserves are estimated to be 1.2×10^8 tons (0.29 to 1.0×10^7 quads). Because reserves are large and consumption small, there has been no need to prospect for new lithium deposits. Lithium deposits are widely scattered. The major producers are the United States, Zimbabwe, the Soviet Union, Canada, and China. Lithium could also be obtained from the oceans if there was a sufficient demand.

Deuterium is available in practically unlimited quantities in the oceans, in the form of D_{2O} or HDO. Natural oceanic water contains 16.687 ppm deuterium. It is estimated the oceans contain 4.6×10^{13} tons of deuterium, providing an energy source of 15×10^{12} quads.

Because most countries have access to the oceans, a supply of lithium and deuterium for nuclear fusion is readily available. When fusion becomes available, a safe and unlimited supply of energy will be available to the world.

In some types of reactors beryllium is the required breeder material. Estimates indicate that beryllium is available in sufficient quantities to satisfy all demands. The average abundance of beryllium in the earth's crust is 6 ppm. It is estimated that a beryllium reactor consumes about 0.09 the amount of fuel required for a lithium reactor. No radiation is emitted in nuclear fusion, regardless of whether lithium, deuterium, or beryllium is used.

Milling

After mining, the uranium ore undergoes a process known as milling (Thomas C. Hallocher and James J. MacKenzie, "Radiation Hazards Associated with Uranium Mill Operation," in Union of Concerned Scientists, *The Nuclear Fuel Cycle* [Cambridge, MA: MIT, 1975], pp. 41–69). The ore is finely ground and most of the uranium and thorium is chemically extracted. The fine residue that remains, known as tailings, still contains most of the original radioactivity—typically 700 picocuries (one trillionth of a curie) per gram—more than 100 times that found in ordinary rocks. In addition, radium is also found in liquefied water from the extraction process and in a mudlike by-product known as slime. These solutions are usually deposited in huge piles, or after dilution to appropriate radiation levels in nearby streams. Most tailings in the western United States are found in 35 large piles on mill sites. From 1946 to 1968 it is estimated that at least 100 million tons of tailings were produced.

Occupational Hazards to Uranium Mill Workers

There is a growing body of evidence suggesting that exposure to uranium tailings is more hazardous than previously suspected. A study of 662 mill workers that began in 1950 showed no change in the mortality rate until 1966, and in 1967 four deaths due to lymphatic cancer were reported. The lymphatic dose from thorium 230 is about 20 times greater than that for uranium 238 and uranium 234 together. These three isotopes are found in about equal amounts in the decay of uranium.

In the four cancer deaths the mill workers were exposed to thorium concentrations of about 4×10^{-11} μCi/C/m^3 for periods of 5 to 13 years. The International Commission on Radiological Protection (ICRP) has recommended that the maximum permissible concentration for thorium 230 be 4×10^{-12} μCi/C/m^3. This standard may not give sufficient health protection to mill workers.

Radium in Streams

By the late 1950s it was recognized that a number of streams near uranium mill sites had been badly polluted with radium and that water for human consumption and irrigation was endangered. To illustrate, at a uranium mill site in southwest Colorado, the tailings were dumped into the Animas River. By 1958 damage to the river could be traced 50 miles downstream. Radioactivity, principally radium, was 500 times the normal amount found in the stream and 8 times the Radiation

Protection Guide (RPG) limits (the limit is 3 pCi of Ra 226 per liter). About 30,000 people lived along the Animas, largely in Aztec and Farmington, New Mexico, and used the river for drinking water and irrigation. The drinking water frequently exceeded the RPG limits. As the minerals entered the food chains, the flora and fauna were found to have concentrations of radium 10^2 to 10^4 times that found in the open water. Grasses and alfalfa irrigated with the river water concentrated radium on the order of tenfold, which was then passed on to the livestock.

By 1959 the U.S. Public Health Service recognized the seriousness of the problem, and new regulations by the Atomic Energy Commission stopped or diminished the dumping of tailings in streams. This solved part of the problem, but most of the great piles of tailings remain unstabilized and continue to be eroded by wind and water.

Radiation Hazards from Mill Tailings

In the milling of uranium ore more than 99 percent is waste, so near the mill sites huge piles of tailing accumulated. For many years the danger from radioactivity was not recognized. Because of the sheer volume, the Environmental Protection Agency believes that mill tailings may present the greatest environmental impact of all waste forms over the years. The extraction of a few pounds of uranium to fuel a single 1,000-megawatt light water reactor for a year requires the generation of at least 106 tons of tailings.

In 1958 the AEC discovered that radon emanating from the piles created a high level of airborne radioactivity in the immediate vicinity of the tailings. Because there was a high level of gamma rays near the surface of the tailings, access to the tailings was restricted in most areas. However, the public could remove tailings for use as fills or other purposes until 1969. The greatest danger to individuals near tailings comes from radon daughters and gamma rays.

For many years the fine-ground tailings were used for many purposes by the public. The use of tailings in Grand Junction, Colorado, illustrates the early use of the material. The mill of the Climax Uranium Company provided a continually new supply of material. The tailings were used as fill at a number of construction sites, as a substitute for sand in concrete and mortar, and for other uses. The tailings had been removed from the mill and used with the knowledge and permission of the AEC. In 1966 Colorado public health officials found abnormally high readings of gamma radiation and radon daughters in certain buildings. Investigation revealed that $2–3 \times 10^5$ tons of tailings had been used in construction in Grand Junction for 15 years prior to 1966 and that about 3,000 homes and buildings had tailings under and around their foundations. Records had not been kept as to where the tailings had been deported, so their location was determined by gamma ray surveys.

The use of tailings leads to two radiological hazards. The gamma rays can penetrate concrete and masonry walls, giving whole-body exposure inside. Radon also moves through concrete materials and the radon daughters cause an airborne source of radiation indoors.

In 1973 more than 5,000 indoor sites in Colorado were surveyed to secure gamma ray data (see Table 3). The median gamma level is about 13 Mrad/hour, and background is taken to be 10 Mrad/hour over an 8,000-hour year yield. The gamma ray background of 10 Mrad/hour over an 8,000-hour year yields 80 Mrad/year whole-body dose out of a total background in Colorado of perhaps 150 Mrad/year. The median additional dose due to tailings is about 3 Mrad/hour and represents maximally (assuming constant exposure) only about 24 Mrad/year. About 20 percent of the sites exhibit values in excess of 20 Mrad/hour. In those 5 percent of the sites where the additional exposure from tailings is greater than 50 Mrad/hour, the average additional dose became about 450 Mrad/hour or about 3 times total background. The Surgeon General's guidelines of 1970 recommend remedial action at greater than 100 Mrad/hour above background for gamma rays and suggest remedial action for 5–100 Mrad/hour above background.

The hazard from radon daughters is perhaps greater than that from gamma rays (see Table 4). If we consider an integrated value above background of 0.025 WL over 1 year, continuous exposure would include about 50 working month units of 170 hours each year for a resident. This amounts to 1.25 WLM accumulation/year, and to a living dose of about 2.5 rad/year. Over 30 years this amounts to nearly 75 rad to the lungs. This is about one third of the doubling dose for lung cancer and potentially involves several thousand persons. Levels in about 5 percent of the buildings associated with tailings would be greater than 10 times this dose. (Data from *The Nuclear Fuel Cycle,* 1975.)

The solution to the mill tailings problem is not an easy one. The ultimate solution to this problem is the complete removal of the tailings. Because of the massive amount of material and because it is scattered over wide areas of the southwestern United States, the cost could be as much as $100 million. Several measures can be taken to minimize the radium hazard from the tailings. The simplest procedure is to flatten the piles and cover them with earth. One objection here is that the half-life of radium 226 is 1,630 years. Thus, the radium is likely to outlive the protection of a thin earth covering. The best procedure would be to return the tailings to the mill and recover the radium content. The tailings could then be mixed with cement to ensure the

TABLE 3

Indoor Gamma Ray Survey for Buildings Associated with Mill Tailings in Colorado[a]

SCINTILLATION COUNTER READING	CORRECTED RAY LEVELS[b] (Mrad/hour)	NUMBER OF BUILDINGS	PERCENT OF BUILDINGS
0–25	0–14.9	3,766	69
25–40	14.9–23.2	715	13
40–60	23.2–34.5	342	6
60–100	34.5–57	272	5
100–200	57–113	283	5
200–300	113–169	41	0.8
300–500	169–281	24	0.4
500–1,000	281–561	4	0.1
>1,000	>561	1[c]	—

[a]Survey of 5,448 buildings in Colorado in late 1973.

[b]Corrected ray level = 0.568 + 0.9 in Mrad/hour and includes gamma ray background estimated at 10 Mrad/hour. Median level, 13 Mrad/hour; average 23 Mrad/hour.

[c]Factory.

Source: *The Nuclear Fuel Cycle* (Cambridge, MA: MIT Press, 1975), pp. 56–57.

TABLE 4

Integrated Indoor Radon Survey for Buildings Associated with Mill Tailings in Grand Junction, Colorado[a]

WL RANGE[b]	NUMBER OF BUILDINGS	PERCENT OF BUILDINGS	APPROXIMATE MAXIMUM IONS TO LUNGS (rad/year)
0–0.005	62	10.0	0–0.5
0.005–0.01	133	21.5	0.5–1.0
0.01–0.025	170	27.5	1.0–2.5
0.025–0.05	118	19.0	2.5–5.0
0.05–0.10	74	12.0	5.0–10
0.1–0.2	31	5.0	10–20
0.2–0.4	18[c]	3.0	20–40
0.4–0.6	5[c]	0.8	40–60
0.6–0.8	2[c]	0.3	60–80
0.8–1.0	3[c]	0.5	80–100
1.0–1.5	2	0.3	150–
1.5–2.0	1	—	—

[a]Survey of 1971–1972 on 619 buildings.

[b]Includes background estimates of 0.005 WL. 1 WL of radon daughters delivers about 100 rad/year to lungs under constant exposure. Median level, 0.021 WL; average 0.057 WL. Integrated values are generally from data over six locations within each structure.

[c]Values at individual locations frequently were in the range of 0.5–1.5 WL.

Source: *The Nuclear Fuel Cycle* (Cambridge, MA: MIT Press, 1975), pp. 54–55.

entrapment of radon. The most sophisticated solution would be to chemically recover the radium, incorporate it with the reactor fuel, and finally destroy it by transmutation and fission.

Conversion

In the milling process chemicals are used to convert and purify the uranium ore with a semirefined uranium oxide (U_3O_8), known as yellowcake. Historically, the size of the uranium industry has been measured by the amount of uranium concentrate produced, not by the ore produced. This is because only about 0.002 percent of the ore becomes uranium oxide. In 1960, 8 million tons of ore produced 17,637 tons of U_3O_8. The quantity of waste rock that has many radioactive materials presents a major environmental problem.

The second stage in the conversion process is to change U_3O_8 to UF_6. In this process there are small radioactive water and air emissions from the plant. The Environmental Protection Agency has indicated that conversion plant emissions produce an average dose of much less than 1 millrem/year to individuals within 50 miles of a plant. The five UF_6 conversion plants in the United States are located at Metropolis, Illinois; Sequoyah, Oklahoma; Barnwell, South Carolina; Apollo, Pennsylvania; and West Valley, New York. The conversion of U_3O_8 to UF_6 has little possible environmental contamination.

Enrichment

The next step in the nuclear fuel cycle is the enrichment of the hexafluoride $UF_{6,}$ which is a gas at conditions near room temperature and pressure. The enrichment is accomplished by a gaseous diffusion process. The UF_6 is forced through about 1,700 barriers in which the U 235 concentration is increased from the natural 0.7 percent to a level of 3 to 4 percent, the balance being U 238. The amount of U 235 produced is determined by technical and economic considerations. In this process large quantities of heat, water, and electricity are required. The enriched fuel for one year's operation of a 1,000-million-watt (MW) light water reactor (LWR) requires about 11 billion gallons of water and 310 million kilowatt-hours of electricity. This is about 4 percent of the energy generated in a plant.

The large amount of cooling water required is the major environmental impact in the enrichment process. Cooling towers are required to reduce the temperature of the water. Other environmental impacts are minimal. The radiation dose to an individual living at the plant boundary would be only about 2 mrem/year.

The guaranteed supply of enriched materials is of critical importance to nations that rely on nuclear power as their major source of electricity. In addition, uranium enriched to very high levels can be used to construct nuclear weapons. The development of enrichment services entails the construction of expensive plants using classified technology, and for plants using the diffusion process, considerable supplies of water and energy. The enrichment stage of the nuclear fuel cycle accounts for nearly 50 percent of the total cost, excluding costs of the nuclear reactor; the other major cost is for the raw material, yellowcake (U_3O_8). The costs associated with the other stages of the fuel cycle are all relatively minor compared with these two components.

Fuel Fabrication

The final step before the fuel is used in the nuclear reactor takes place when the enriched UF_6 is converted into uranium dioxide (UO_2). The UO_2 is formed into small ceramic pellets and enclosed in a thin tube made of a suitable material, usually zircaloy, to form fuel rods. The fuel rods are then transported to the power facility in bundles called fuel assemblies. The number and arrangement of the fuel rods in the fuel assemblies are determined by the specifications of the reactor core design.

The fabrication of fuel elements is a well established technology and unlike the enrichment process, fuel element fabrication plants operate on a commercial basis. In fuel fabrication plants extensive tests and inspections are carried out during all parts of the operation to ensure safety.

The Environmental Protection Agency has indicated that this process has little or no impact on the environment. At the plant boundary of the typical facility, an individual might receive a maximum of 10 mrem/year in the lungs from normal breathing, but the average dose within 50 miles of the plant would be less than 0.1 mrem/year.

Energy Production

One of the most controversial issues in the use of nuclear fuels to produce energy is the fission of atoms in the nuclear reactor. There has always been apprehension about the safety of nuclear reactors, and this controversy has grown with the Three Mile Island and Chernobyl accidents. In the fission process heat is created to produce steam. After the steam is generated, conventional procedures are used to convert the steam into electricity.

Nuclear Fission

An atom is the smallest particle into which an element can be divided chemically. It consists of positively charged particles called protons and an equal number of negatively charged electrons. In addition, there are a number of neutral particles called neutrons. The number of protons in the nucleus determines the chemical properties of the element. Uranium has 92 protons. Atoms of the same element, however, may have a different number of neutrons. These are known as isotopes. For example, uranium 238 has 146 neutrons, and uranium 235 has 143 neutrons. This means that the chemical properties of these two isotopes are identical but the nuclear properties are fundamentally different.

Nuclear fission occurs when certain elements are split by neutron bombardment. Of all elements, uranium is the only fissionable one that occurs in large quantities in nature. Thorium is also fissionable, but occurs only in small quantities in nature. The uranium atom has a number of isotopes, only one, uranium 235, is fissionable. Uranium 238 is nonfissionable but can be converted into uranium 235 when placed inside a nuclear reactor.

When uranium 235 absorbs an additional neutron, it will likely split, or fission, into two smaller nuclei, resulting in the release of a large quantity of energy and the emission of two or three neutrons. The emitted neutrons are now available for absorption by other fissionable nuclei. In this manner a fission reaction is made self-sustaining if the neutrons emitted by one reaction trigger a subsequent fission reaction. Not all neutrons emitted in the fission process initiate subsequent fission. Some neutrons escape from the area of reaction, some are absorbed, and some are captured by heavy nuclei without producing fissioning. The rate of fissions is controlled with moderators, substances used to slow down neutrons in a nuclear reactor. If the emitted neutrons were not controlled, the rate of fissioning and consequent heat output would increase exponentially, resulting in the ultimate explosion of the reactor.

Nuclear Power Reactor Systems

Reactors are usually classified as burners, converters, or breeders depending upon the amount of new fissile material formed during the reactor's operation. When there is greater consumption than production of fissile material in the reactor, it is called a burner. When the ratio approaches 1, it is known as a converter, and when the conversion ratio exceeds 1, the reactor is called a breeder.

There are five types of reactors already in commercial use or at the development stage:

1. The light water reactor (LWR), which uses ordinary water as a moderator and coolant, now provides from 85 to 90 percent of the reactor capacity of the world. This reactor uses U 235 as a fuel and thus requires the availability of uranium enrichment facilities. Two variations of this reactor have been produced—the pressurized water reactor (PWR) and the boiling water reactor (BWR). In the BWR the water heats until it becomes steam to be used directly to drive the turbine generator to produce electricity. In contrast, in the PWR the heat is transferred to a secondary system for the generation of steam.
2. The heavy water reactor (HWR) uses natural uranium as a fuel and heavy water. The pressurized heavy water reactor (PHWR) uses heavy water as both moderator and coolant. As with the PWR, the coolant produces steam to turn the turbines. The steam-generating heavy water reactor (SGHWR) uses heavy water as a moderator and light water as a coolant. Similar to the BWR, water heats to produce steam in the generating process.
3. The gas-coolant reactor (magnox) uses pressurized carbon dioxide and natural uranium fuel rods encased in "magnox," a magnesium alloy, and a graphite-moderated core. An advanced gas-cooled reactor (AGR), still in the experimental stage, will employ enriched uranium, U 235, in the form of uranium dioxide. In this process it is expected that a higher proportion of the uranium 235 will be consumed, thus enabling the reactor to operate at a higher temperature. This reactor is now being tested in the United Kingdom.
4. The high-temperature gas-cooled reactor (HTGR) uses three fuels—highly enriched uranium 235, uranium 233, and thorium 232. The mix of fuels changes over time—the initial fuel load consists of uranium 235 and thorium 232. In subsequent loadings recycled uranium 233 replaces highly enriched uranium. Graphite is used as a moderator. The coolant is helium. The thermal efficiency is high because the coolant gas allows much higher temperatures than other coolants. Only a few HTGRs are in operation, and they have yet to demonstrate economic feasibility.
5. The light metal fast breeder reactor (LMFBR) has been at the developmental stage for many years, but its use has been

impeded by technological and economic restraints. The LMFBR fuel rods contain a mixture of about 20 percent plutonium with depleted uranium dioxide, and a blanket of rods containing depleted uranium dioxide surrounds the core. The neutrons released by the reactor fuel convert U 238 to fissile material (plutonium 239) faster than the fissile material in the fuel is converted to nonfissile material. Hence, the term "breeder" reactor is appropriate. The initial loading can use either plutonium recovered from the spent light water reactors or enriched uranium. Subsequent loadings use plutonium produced in the LMFBR itself. The United States was an early leader in breeder reactors but has reduced its efforts. Some Western European nations have continued development. The first commercial 1,200 MW LMFBR plant was put in operation at Crey Malville, France, in 1985. Smaller plants operate in the United Kingdom, West Germany, and the Soviet Union.

Environmental Considerations

In all nuclear power plants, radioactive materials are produced in the nuclear reactor. The fissioning of the uranium and the neutron activation of the coolant produces many radioisotopes. These radioactive isotopes, particularly the gaseous ones, can escape from the fuel rods through pinhole defects to contaminate the coolant water or gases. Technical specifications limit the amount and rate of release of these radioactive materials into the environment. Constant monitoring of the facilities occurs. The Environmental Protection Agency has estimated that the health effect of one year's operation of a 1,000 MW plant would be responsible for a total of only about 0.001 mrem/year. It is thus evident that the environmental contamination by radioisotopes from an operating nuclear plant that is continually monitored is minimal.

A major concern relates to the cooling system. If the coolant were to stop passing through the reactor core, the fission reactions would stop, for the moderating influence of the coolant would end. At the same time, the heat from the radioactive decay of the fission products could result in the melting of the fuel, and if continued would lead to a break in the pressure vessel and containment system. At this point the radioactive gases would escape violently into the environment. The hazard to the nearby, and possibly even distant, population could be great. This could have occurred at the Three Mile Island accident but was prevented. In contrast, at Chernobyl gases did explode into the atmosphere. These accidents forced a reevaluation of the use of nuclear fuels to produce electricity.

Trends of Nuclear Power Production

The development of nuclear power has been spectacular since 1960. The number of operable nuclear reactors has increased from 66 in 1970 to 234 in 1980 to 398 in 1986 (see Table 5), and net capacity of installed megawatts grew during the same periods from 15,471 to 132,782 to 290,585 (see Table 6). In 1986 nuclear power accounted for about 4 percent of the world's primary energy consumption.

The development of nuclear power has not been evenly distributed over the world. More than 90 percent of nuclear energy production is concentrated in the highly industrial areas. Of the total installed capacity of 290,585 megawatts in 1986, Europe had 36 percent, Anglo-America 33 percent, and the Soviet Union 15 percent, or 84 percent of the world's total.

Fuel Reprocessing

The fuel cells in a reactor must be replaced when they are spent. Because the spent rods are highly radioactive, the normal practice is to store them about a year underwater, during which time the most intense short-lived and intermediate half-life radioactive fission products have reduced radioactivity. The rods are then transported to a reprocessing plant to separate residual uranium and plutonium from the fission products. The recovered uranium, which could still contain around 1 percent U 235, can be converted into uranium hexafluoride (UF_6) for subsequent reenrichment. The recovered plutonium can be converted into plutonium dioxide for subsequent use in mixed oxide fuel elements that are a blend of uranium and plutonium dioxides. The fission products containing the radioactive wastes must be treated and disposed of.

Although only Marcoule, France, The Hague, Netherlands, and Windscale in the United Kingdom have industrial-scale reprocessing plants in operation, other countries have laboratory and pilot plants. The environmental impact of the industrial-scale plants varies considerably. Marcoule, which has tight environmental controls because it discharges into the Rhine River, is the cleanest. The other two plants discharge their waste into the sea. Windscale is a heavy environmental polluter. A large part of the radioactive material it releases does not come from reprocessing, but from the corrosion of fuel cans stored while awaiting treatment. Between 1975 and 1979 Windscale's discharges to the sea were responsible for more than 3.5 times more beta

TABLE 5

Number of Operable Reactors

LOCATION	1970	1975	1980	1986
United States	13	52	68	100
Soviet Union	11	17	26	53
France	3	6	17	49
United Kingdom	25	28	32	38
Japan	4	12	23	34
Germany, Fed. Rep.	1	4	10	19
Canada	1	5	9	18
Sweden	0	5	8	12
Spain	1	3	3	8
Czechoslovakia	0	1	3	8
Belgium	0	1	3	7
India	2	3	4	6
Korea, Rep.	0	0	1	6
Taiwan	0	0	2	6
Bulgaria	0	2	3	5
Switzerland	1	3	4	5
Germany, Dem. Rep.	1	3	5	5
Finland	0	0	4	4
Hungary	0	0	0	3
Italy	2	2	3	3
Netherlands	1	2	2	2
Argentina	0	1	2	2
South Africa	0	0	0	2
Pakistan	0	1	1	1
Yugoslavia	0	0	0	1
Brazil	0	0	1	1
TOTAL	66	151	234	398

activity and 75 times more alpha activity for each gigawatt year of electricity produced than were The Hague's during the same period. Windscale has since improved but is still polluting more than The Hague. New plant design can reduce the release of radioactive materials considerably. At the present time less than 10 percent of spent fuel is being reprocessed, and even by the early 1990s when some large-scale commercial plants could be operational, less than half of the required reprocessing capacity will be available. Delays in reprocessing are related not only to technical problems but also to political and institutional problems of nonproliferation.

Uranium enrichment and reprocessing of spent fuel are considered the most sensitive elements of the fuel cycle from the nonproliferation point of view. Although initially the recovered plutonium is not weapons grade, it is technically possible to upgrade it in quality for the production of weapons. Transfer of technology and international cooperation in this area has essentially been lacking.

TABLE 6

Net Capacity (megawatts) of Reactors

LOCATION	1970	1975	1980	1986
United States	6,411	36,475	51,176	85,177
France	1,190	2,478	12,468	44,873
Soviet Union	1,438	4,438	11,428	43,105
Japan	1,258	6,292	15,011	24,754
Germany, Fed. Rep.	328	2,744	8,523	18,944
United Kingdom	3,402	4,492	6,956	11,748
Canada	22	2,078	5,150	10,989
Sweden	0	3,130	5,515	9,650
Spain	153	1,073	1,073	5,668
Belgium	0	393	1,656	5,450
Taiwan	0	0	1,208	4,884
Korea, Rep.	0	0	564	4,480
Switzerland	350	1,020	1,940	2,930
Czechoslovakia	0	104	864	2,874
Bulgaria	0	810	1,215	2,713
Finland	0	0	2,266	2,310
South Africa	0	0	0	1,840
Germany, Dem. Rep.	70	886	1,702	1,702
Italy	397	397	1,270	1,282
Hungary	0	0	0	1,230
India	400	602	800	1,164
Argentina	0	345	945	935
Yugoslavia	0	0	0	632
Brazil	0	0	626	626
Netherlands	52	497	497	500
Pakistan	0	125	125	125
TOTAL	15,471	68,379	132,978	290,585

The Environmental Protection Agency has attempted to predict the health effects at a reprocessing plant. It is estimated that a plant with a capacity of 5 metric tons per day servicing 45 nuclear plants of 1,000-MW size would emit long-lived radioactive gases such as krypton 85 and tritium at a rate that would produce about 2.5 health effects annually in the United States and about 100 health effects annually worldwide. If the processing plants were increased in number to service 1,000 nuclear power plants, the health effects would increase to about 200 annually worldwide, of which there would be 130 cancer deaths.

Radioactive Waste

Classification of Radioactive Waste

Radioactive waste has been legally defined as belonging to one of four categories: high-level and spent nuclear fuel, low-level waste, transuranic waste, and mill tailings.

High-Level Waste (HLW). The Nuclear Waste Policy Act of 1982 defines high-level waste as "the highly radioactive material resulting from the reprocessing of spent nuclear fuel, including liquid waste produced directly in reprocessing and any solid material derived from such liquid waste that contains fission products in sufficient concentrations; and other highly radioactive material that the Nuclear Regulatory Commission (NRC) by rule requires permanent isolation." Spent nuclear fuel is defined by the act as "fuel that has been withdrawn from a nuclear reactor following irradiation, the constituent elements of which have not been separated by reprocessing." This waste containing fission product and transuranics is highly radioactive and requires massive shielding during handling. Some of these waste products have a very long half-life (thousands of years) and therefore present a problem of storage never before faced by a society. The Nuclear Waste Policy Act of 1982 gives the responsibility for the ultimate disposal of all high-level waste to the federal government.

Low-Level Waste (LLW). The Low-Level Waste Policy Act of 1980 defines low-level waste as "radioactive waste not classified as high level radioactive waste, transuranic waste, spent nuclear fuel or mill tailings." The act made the states responsible for the disposal of low-level waste. It permits states to enter into compacts for the development of regional disposal facilities. These regional compacts, involving several states, may exclude waste from other compact regions. Low-level waste is produced by a large variety of industries and institutions that use radioactive material. It thus comes in fairly large quantities and in different forms.

To solve the problem of handling different types of low-level waste, the NRC has developed an LLW classification scheme dividing the waste into three classes: A,B, and C. The class into which the LLW falls is determined by the concentration of various radionuclides. Class A contains the lowest concentration of radionuclides and must meet minimum standards for stability, such as:

Storage in cardboard or fiberboard is not acceptable

Solid wastes must contain as little liquid as possible

Waste cannot be pyrophoric, explosive, or highly reactive or capable of generating gases

Hazardous material must be removed from the waste to the greatest extent possible

Class A waste is considered relatively unstable compared to Class B and C. It may thus decompose and cause slumping of materials at the disposal site. This could lead to the eventual failure of the disposal site to contain the waste material. Because this type of waste consists of low concentrations and short-lived radioactive waste materials, there is little or no danger to the public. The stability requirement of Class A waste distinguishes it from Classes B and C.

Classes B and C must meet the same disposal standards as Class A waste but in addition must meet more rigorous requirements concerning waste form to ensure that the waste will remain stable after disposal. These additional requirements include:

> The depository must have structural stability. This can be accomplished by placing waste into special containers.
>
> All liquids must be converted to as solid a state as possible.
>
> All voids in the disposal site must be filled so that there can be no movement of materials.

The development of disposal sites for Classes B and C waste may require engineered barriers to prevent leakage of radioactive wastes. If the waste is highly radioactive, disposal must normally be at a greater depth. This becomes a major problem when the water table is near the surface. If this occurs, special arrangements may be needed to ship the waste to another region for disposal.

Transuranic Waste (TRU). Transuranic waste is defined as "waste material containing radionuclides with an atomic number greater than 92, which are excluded from shallow land burial by the Federal government" (Northwest Interstate Low-Level Radioactive Waste Management Compact, 1983). Originally, waste containing these elements with a concentration greater then 10 nano (10^{-9}) curies per gram of waste was considered TRU waste. The NRC raised this limit by an order of magnitude to 100 nanocuries per gram of waste. TRU is largely generated by the defense activities of the federal government, and permanent disposal rests with the federal government. Because TRU radionuclides have a long half-life, they must be isolated in a satisfactory geologically safe repository similar to that for high level waste.

Mill Tailings. Mill tailings are defined as "the remaining portion of the metal bearing ore after some or all of the material, such as uranium, has been extracted, or other wastes produced by the extraction or concentration of uranium or thorium from any ore processed primarily for its source material content" (Uranium Mill Tailings

Radiation Control Act of 1978). Mill tailings are produced primarily from the extraction of uranium from ore for nuclear fuel. These wastes are produced in large quantities, but the radioactivity is low. The great danger lies in the daughter products of uranium and thorium, such as radon and radium. Some of these have very long lifetimes. Mill tailings are usually disposed of near the mining operations and because of the huge volume present a difficult safety problem.

Reactor Nuclear Waste

The final step in the nuclear fuel cycle is management and disposal of radioactive waste, some of which may have a radioactive half-life of tens of thousands of years. Some forms of waste are generated at nearly every stage of the nuclear fuel cycle, from the tailings at the mill site to the high-level fuel-reprocessing operations. This problem has attracted much attention, and public fears have become so pervasive that the entire nuclear energy industry has been threatened.

In the process of producing electricity, for every metric ton of nuclear fuel in an initial load, 24 kg of uranium 238 and 25 kg of uranium 235 are consumed in a three-year period. In the fusion process the enriched uranium 235 is reduced from 3.3 percent to 0.8 percent of the total. The consumption of 1 ton of nuclear fuel produces 800 million kilowatt-hours of electrical energy.

In the process of producing electricity, 35 kg of radioactive waste products are produced in the form of isotopes in the following amounts: 8.9 kg of plutonium, 4.6 kg of uranium 236, 0.5 kg of neptunium 237, 0.12 kg of americium 243, and 0.04 kg of curium 244. Because only 25 kg of uranium 235 are consumed and a fifth of that amount is converted into uranium 236 and neptunium 237, it is calculated that only 60 percent of the energy produced comes from uranium 235.

Management of Nuclear Waste

Although the nuclear power industry is more than 30 years old in the United States, all nuclear waste has been placed in temporary depositories. These depositories contain more than 99.8 percent of the non-gaseous fission products produced in the reactor. At present, these liquid wastes are usually concentrated by evaporation and stored as an aqueous nitric acid solution in high-integrity, stainless steel tanks. There is now general agreement in the nuclear industry that these wastes should be solidified and then placed in a permanent depository.

With the development of the nuclear industry, spent fuel inventories have grown rapidly. Statistics are only available for the market economies and are missing for the nonmarket ones, primarily the Soviet Union. For the market economies the cumulative metric tons of heavy metals in the spent fuel inventories increased from 6,205 in 1970 to 16,714 in 1975, to 34,913 in 1980, to 76,800 in 1986. Of this total the United Kingdom had 27,400 metric tons in 1986; the United States, 14,000; France, 10,800; Canada, 8,700; Japan, 4,300; and West Germany, 2,500 tons.

Radioactive Waste from Medical Uses

Soon after the discovery of radium, people recognized that it had potential medical uses. Today most health care facilities have some type of radiological equipment. Because these treatments use radioactive materials, there has been concern in recent years that many patients have received doses of radiation that are unnecessarily high. As a result there have been attempts to reduce the amount of radiation. This has been accomplished by using image-intensifying screens for X-ray work, a more careful use of radiation shielding in the use of X-rays and radiation therapy, and the availability of a greater variety of lower-dose-producing radionuclides for medical applications. Radiological standards have been established by federal, state, and even local agencies to minimize the risk of personal health hazards. All radiological services are required to develop policies and procedures for safe use of radiation treatment.

The use of radiation in medical diagnosis and treatment is the largest source of radiation exposure for the general public. In the United States the average genetically significant dose (GSD) is about 55 mrem per year. Dental radiation is estimated to contribute an additional 0.15 mrem, and therapeutic radiation about another 0.10 mrem.

Radiological treatments create low-level radioactive waste. Because of the magnitude of the treatment in the many health centers and hospitals, the quantity of low-level waste has increased greatly in recent years. Its safe disposal has become so important that national legislation has been passed, and national policies are gradually being developed for its disposal.

Types of Disposal of Nuclear Waste

Numerous studies of the types of permanent disposal of nuclear wastes have been conducted. No universal agreement has been reached. One of the major problems in choosing and identifying suitable sites is the vocal objections of the general public. Although there is universal recognition that the waste products must be disposed of, there is also the viewpoint, "not in my backyard." The problem is thus not only a technical one but also one with wide political ramifications.

Land Disposal

The details of land disposal of radioactive waste are still not definite, but most research indicates that the following technique is most promising. The waste would be placed in some type of borosilicate glass cylinders about 300 centimeters long and 30 centimeters in diameter. These containers would then be placed in stainless steel containers. It would require 10 canisters each year to contain the waste from a single 1,000-megawatt nuclear power plant. Each of the canisters would be lowered to a depth of at least 600 meters about 10 meters apart in order to dissipate the heat generated by the decay of the radioactive waste. The problem of excessive heat could be controlled by delaying the burial by about one decade. It is estimated that about one-half square kilometer of space would be needed annually to store the radioactive waste of the U.S. industry.

For land disposal, the geology of the area is a critical factor. It must be tectonically stable for thousands of years. Volcanic activity must also be absent. Many scientists believe that salt deposits would provide the best sites. In such deposits water would migrate toward the canisters, reducing temperature considerably. Typical salt deposits contain about 0.5 percent water trapped between the salt particles. The reduction of temperature is critical, for glass devitrifies (crystallizes and becomes brittle) at temperatures higher than 700°C. In time it is assumed that the stainless steel canisters will decay, leaving the glass containers in contact with the salt.

Although complete risk cannot be eliminated in the burial of radioactive wastes, the danger must be placed in a proper perspective. A public fear is that the canisters will ultimately leak and the groundwater will contaminate food and drinking sources. Because the canisters would be placed in impervious layers of rock, the typical rate of flow of water would be less than 30 centimeters per day. At this rate it would take about 1,000 years for the water to migrate 100 kilometers (62.5 miles). During this period the decay of the waste makes it far less dangerous. For example, cesium 137 decays to barium 137 in a period of 400 years. During this period the total gamma ray hazard falls by more than four orders of magnitude.

There is also a fear that the waste material would reach the surface because of erosion. This is an unfounded fear. When the waste is buried more than 600 meters deep, millions of years would be required to erode it to the earth's surface.

Another fear is that social and political conditions are not sufficiently stable to guarantee safety. Monitoring excessive ionizing radiation would require a minimum of personnel. Because of the radiational effects of the waste materials, radical groups will not disturb the permanent deposits.

Incineration

A large percentage of the low-level radioactive waste originating from various parts of the fuel cycle, from nuclear research institutions, and from industrial and medical activities is material that is combustible. These combustible wastes represent at least 50 percent of the low-level solid wastes and can account for almost 80 percent of the total volume produced. These wastes are produced in very large volume with relatively low radioactivity.

The most efficient way to treat these wastes is by incineration. This process reduces the bulk by converting the organic material into ashes and residues that can be compacted into forms suitable for safe storage and disposal. In special high-temperature incinerators the wastes are converted into slags that safely contain the radionuclides in highly insoluble form, and thus are stored.

In incineration the net volume may be reduced by a factor of nearly 100 and the weight by a factor of 20. The reduction depends on the type of incineration and on the secondary wastes produced during the operation of the incinerator. Nevertheless, the overall volume reduction achieved by incineration is higher than that of other potential types of treatment.

There are a number of technical problems that must be considered in the incineration process. These include the nature of the waste, its calorific value, its physical state, its chemical composition (which determines the resulting combustible gases), and its kinds and degrees of contamination. To solve these problems, conventional process engineering and nuclear technology must be combined to arrive at an optimal design and operation of the incinerator system. The goals of the system must consider complete combustion, the control of combustion gases, and radiological protection.

The chemical composition of the waste can lead to noxious combustion gases that must be controlled. These gases can be of a highly corrosive nature requiring special precaution and construction materials. The nature of the contaminants is particularly important if alpha rays are emitted. In these situations special alpha-tight construction principles must be applied.

Although incineration offers an ideal volume-reduction technique for combustible radioactive waste, it is rather difficult to design a universal incinerator that is capable of treating the different types of wastes at equal efficiency and performance. Incineration is suitable for both low-level and intermediate-level wastes. It is not satisfactory for high-level wastes such as those from a nuclear reactor.

Of major concern to the general public has been the control and safety aspects of incineration of radiological waste. Control of the complete incinerator operation is made possible by appropriate temperature, pressure, and flow measurements, which can be grouped together in such a way that a high degree of automated operation can be achieved as well as fully manual operation. All readings are centralized in a control panel that receives information from the various electromechanical and pneumatic components of the installation, such as fans, pumps, burners, and motors.

To prevent accidental fires in the incinerator plant, a clear separation of the different work areas is made. A monitoring system for smoke and fire detection is found in all incinerators. To prevent the escape of radioactive gases into the incinerator and environment in the event of excess pressure in the combustion chamber, the furnace is provided with a relief chimney combined with a calibrated safety valve. Radiological control is provided by the usual stationary radiation control monitor, air sampling devices, and mobile measuring instruments are used in routine contamination surveillance.

Disposal in the Ocean

For many years Britain has practiced the disposal of high level radioactive waste in the sea. Several thousand tons of this waste are deposited in the sea each year. After international consultation, a site was chosen several hundred miles off Lands End, the westernmost point of England. At this site the water is deep, there are no strong upcurrents, and fishing is not pursued. The waste is placed in containers that break up after several years. The waste is then diluted by the vast quantities of the ocean's waters. Computer models indicate that no hazard exists to life in the ocean or on nearby land bodies. There have been vociferous protests from conservationists and environmentalists, but the practice continues.

Most countries, including the United States, have rejected dumping radioactive waste into the sea. When sea dumping occurs, there is a complete loss of control of the waste material. There is thus the fear that the waste material could create a hazardous environment.

Sub-Seabed Disposal (SSD)

Among the considerations for the disposal of high-level waste has been its deposition beneath the seabed. From a technical and legal viewpoint, SSD is not dumping, but emplacement beneath the seafloor. The concept of SSD was first proposed by scientists in the United States, but since 1976 an international Seabed Working Group (SWG) has met under the aegis of the Nuclear Energy Association. The SWG now has representation from nine nations. These representatives account for about three quarters of the world's installed nuclear capacity.

The most recent studies propose that the clays of the deep seabed be used as a geological medium for isolating high-level waste. The waste, packed in metal canisters, would be buried in the sediment 60 to 150 feet below the seafloor. Studies completed support the initial findings that the sediments provide a degree of isolation equivalent to that obtained in land burial. Far from the edges of the continental plates, the ocean floor is seismically stable. Far from the coasts, with their influx of nutrients, the ocean is a desert. This lessens the biological impact of inserting highly radioactive wastes into the seabed. Because the sediments of the seabed are thick having been deposited over millions of years, it is possible to make long-term predictions as to their geologic stability.

The transportation of radioactive waste to the area of repository poses novel engineering problems. Waste, packed in slender canisters 3 feet by 3 inches in diameter, could be deposited in the seabed by a number of techniques. These range from free fall to lowering them into drilled holes. Emplacing waste containers in a controlled fashion from a platform 10,000 to 15,000 feet above the seabed is not simple. The technologies of offshore drilling for oil and gas suggest, however, that no fundamental problems exist. Developing an ensured capacity to retrieve a canister could be more difficult.

The system for transport and emplacement may pose greater difficulties at the landward end than at the repository site. Facilities for marshaling substantial quantities of radioactive waste are unlikely to be welcome tenants in urban ports, and the movement of radioactive waste through highly trafficked waters may have opposition. Military posts may provide the best answer to this problem.

The ultimate answer to whether this system is feasible will likely depend on considerations that lie beyond the technological problems. The issue of feasibility will depend on an assessment of the level of risk with respect to accidents, release rates, potential pathways of released radionuclides to the public, possible harm to the marine environment, and possible harm to man. In addition, feasibility will depend on such issues as political acceptability. The public must accept the system. In the international arena, governmental decisions will also be important. Securing official and public agreement will be complicated by the need to make comparisons between land and sub-seabed options. At the present time it is an option that is not being pursued by any country.

Nuclear Depositories for High-Level Waste

Since the nuclear industry began in the 1950s, highly radioactive waste has been accumulating from weapons plants, research centers, and nuclear power plants with no permanent storage facility yet completed in most countries. In 1984 the U.S. Department of Energy (DOE) prepared a partial solution by constructing a Waste Isolation Pilot Plant (WIPP) near Carlsbad, New Mexico. This plant will store high-level waste temporarily until a permanent disposal site has been completed. In 1984 the DOE proposed three sites for permanent disposal facilities. These were Hanford, Washington, Hereford, Texas, and the Nevada Atomic Test Site, Yucca Mountain, in the desert of southern Nevada. In spite of opposition, particularly from the Texas and Nevada governors, the Nevada site was chosen.

Carlsbad Waste Isolation Pilot Plant. Since 1984 the DOE has spent $700 million to construct an underground storage depot. Carved out of salt deposits huge storage rooms more than 600 meters (1,970 feet) beneath the surface began receiving nuclear waste late in 1989. The radioactive materials disposed of here will include tools, rags, clothing, and other radioactive debris. The site will be adequate for low-level waste disposal for 25 years. The theory is that within 70 years salt will slowly creep in, encasing the waste in salt and containing the radiation. Federal and state agencies and oversight groups still must review studies of the plant's environmental impact and the means of transporting the radioactive waste to the site.

Yucca Mountain Project. In the desert of southern Nevada in the area of the Yucca Mountain, geologists and engineers are beginning the task of preparing an underground depository to isolate high-level radioactive waste. The estimated cost of the project is $2 billion and the completion date is 2003. Although the initial work has begun on the project, there is much opposition to it. Robert R. Loux, executive director of Nevada's Nuclear Waste Project Office, is opposed to the venture. He believes that Congress favored the repository in southern Nevada because of the very low population of the area. But there is no way to predict future populations. For example, who would have predicted 150 years ago that Las Vegas would have more than 600,000 people today?

Teams of geologists and engineers are attempting to predict whether the area provides a safe depository for high-level waste.

Though some of the waste that would be deposited could be hazardous for millions of years, predictions are being limited to 10,000 years. Beyond this time span, most questions of safety are simply unanswerable. For example, patterns of precipitation could be altered, making present-day analysis of ground and surface water irrelevant.

There has been conflict between the geologists and engineers on the project. Whereas engineers complain that the geologists' predictions reach beyond the limits of their expertise, geologists contend that the engineers sacrifice scientific rigor for adherence to procedures used to ensure only satisfactory present-day engineering practices. At the Yucca Mountain site geologists have clashed with engineers who insisted that all work proceed according to a system called "quality assurance." Under this procedure engineers plan and execute a job, then check to see if the plan was followed precisely in order to guarantee satisfactory work. Geologists in speculating what will happen in the distant future cannot follow such a structural procedure. In the summer of 1988 the 16 geologists from the Denver office of the U.S. Geological Survey complained that quality controls were counterproductive in the pursuit of good scientific investigations.

To illustrate the differences in approaches, in 1988 the engineers drilled temporary holes in the mountain to study structure. The geologists wanted to sample the gases coming from them for gas circulation is critical in that nuclear waste produces radioactive carbon 14 gas. The engineers forced the geologists to stop their investigations because they did not have an approved quality assurance. The geologists indicated that because of bureaucratic opposition, the holes would probably be filled by water, and the opportunity to gather the data, which might reflect unfavorably on the suitability of the site, could be lost. The intellectual clashes between geologists and engineers at the Yucca Mountain site has made it more difficult to predict how the area will be 100 centuries in the future. However, because it is the type of problem never faced by society in the past, a more satisfactory answer may evolve than if there were no differences in viewpoints.

Mankind is faced with one of its most difficult problems. Completely rational decisions may not be possible. With the best advice from scientists and engineers, a permanent depository of high-level nuclear wastes is fundamental to the welfare of all people.

Transportation

A massive amount of material must be transported from one place to another at different times in the processes of the nuclear fuel cycle. It has been estimated that operating a single nuclear facility of 1,000-MW capacity requires truck shipments totaling 16,880 miles of uranium ore on private land, 33,500 miles of shipments of uranium oxide and hexafluoride on public highways, and 2,000 miles of rail shipment of fission products. In this transfer of radioactive materials there is always potential danger to the environment.

The transportation of radioactive material has been regulated since the early 1950s, when concern was expressed for the fogging of photographic film transported near radioactive materials. The early regulation had a twofold purpose—the first was to protect radiosensitive materials and the second was to protect the transport workers and the public at large.

Over the years regulations have developed as radiation protection philosophies and practices have evolved. Because this is a universal problem, the International Atomic Energy Agency (IAEA) has developed international agreements to foster radiation protection principles. The U.S. regulations have evolved according to IAEA recommendations.

Categories of Materials

For transport purposes, radioactive materials are defined as those that spontaneously emit ionizing radiation and have a specific activity exceeding 2uCi/kg. Materials with a lower specific activity are not regulated in transportation.

To facilitate transportation, a number of categories have been devised. The broad spectrum of radioactive materials includes

1. Excepted quantities—very low total radioactivity
2. Low specific activity (LSA)
3. Low-level waste (LLW) type A
4. Low-level waste (LLW) type B
5. Fissile materials—capable of undergoing nuclear criticality

Each category has specific requirements, the stringency being based on the hazard of the material being transported. In all situations the requirements must provide suitable

1. Containment of the material
2. Protection from radiation exposure
3. Rejection of decay heat
4. Prevention of criticality

Excepted Quantities. Excepted quantities present the least radioactive hazard. Three categories of these materials have been devised:

1. Limited quantities—any form of material
2. Instruments and articles—manufactured goods with radioactive materials
3. Manufactured articles of uranium and thorium

These products take many forms and include diagnostic kits, electron tubes, smoke detectors, and samples. The basic requirements are that the radiation level at the surface of the package must not exceed 0.5 mrem/hour, the surface of the package must be free of significant removable contamination, and the materials must be retained in their package under conditions encountered in normal transportation. For uranium and thorium articles, a durable protective sheath must be provided.

Low Specific Activity (LSA) Materials. LSA materials are also of limited potential hazard and have low specific activity. Unlike other categories, radioactivity is not based on total activity of the material but on the specific activity. Five categories of LSA materials have been devised:

1. Uranium and thorium ores and physical and chemical concentrations of these ores
2. Unirradiated natural or depleted uranium and unirradiated natural thorium
3. Tritium oxide in aqueous solutions with a concentration that does not exceed 5.0 mCi/ml
4. Nonradioactive objects externally contaminated with radioactive materials—contamination must be readily dispersible—that when averaged over 1 square mile must not exceed 0.0001 mCi/cm^2 for radionuclides with an A_2 value of not more than 0.05 Ci and must not exceed 0.001 cCi/cm^2 for other nuclides (see Table 7)
5. Materials in which the activity is essentially uniformly distributed and in which the estimated average concentration does not exceed

   0.001 mCi/g for radionuclides with an A_2 value of not more than 0.05 Ci

   0.005 mCi/g for radionuclides with an A_2 value of more than 0.05 Ci but not more than 1 Ci

   0.3 mJCi/g for radionuclides with an A_2 value exceeding 1 Ci

To determine if a given material is LSA or not requires a rather complicated calculation.

TABLE 7

Examples of Radionuclides

RADIONUCLIDE	A_1(Ci)[a]	A_2(Ci)
241Am	8	.008
14C	1,000	60
60Co	7	7
99Mo	100	20
239Pu	2	.002

[a]Because of the wide range of radiotoxicities presented by the radionuclides transported, a system of ranking them has been devised. The system incorporates the metabolic data and dosimetric modeling of the International Commission on Radiological Protection, which is then factored into six series of transportation-related models. The results are two values for every radionuclide: an A_1 value for the "special" form and an A_2 value for the "normal" form. The "special" form refers to radionuclides that are essentially nondispersible, usually because they are contained in high-integrity sealed capsules. Because A_1 values are limited by external radiation, they are generally higher than the A_2 values, which are limited by internal organ considerations. An arbitrary ceiling of 1,000 Ci is placed on the less radiotoxic nuclides. A complete list of radionuclides and their A values is given in Title 49 of the Code of Federal Regulations, Section 173.469 (49 CFR 173.469). Multiples and submultiples of the A values are used throughout the regulations to specify limits because they provide reference values for each radionuclide based on its radioactive properties and the related transportation hazard.

Low-Level Waste (LLW) Type A Materials. Type A materials do not present a serious hazard when transported if allowed to be shipped in packaging designed for normal transport. These quantities are restricted to activities not exceeding the A_1 value for special form materials or the A_2 for normal form materials.

The packaging for type A materials must be capable of withstanding

1. Temperature range of −40°C to 70°C
2. Acceleration, vibration, and resonance that may occur in normal transportation
3. Reduced pressure of 0.25 atm
4. Water spray test simulating exposure to rainfall of 5 cm/hour for 1 hour
5. Free drop of 1.2 meters
6. Corner and rim free drops for certain fiberboard and wooden packagings
7. Compression of 5 times the package weight of 1.3 mg times the vertically projected area in miles
8. Penetration of a hemispherically ended 3.2-cm-diameter steel bar with a mass of 6 kg dropped 1 meter

The design of the package must be designed so that there is

1. No loss or dispersal of the radioactive contents
2. No significant increase in the recorded or calculated radiation levels at the package surface

Low-Level Waste (LLW) Type B Materials. Type B materials present a significant hazard if the contents are released. In addition to meeting all of the type A packaging criteria, type B materials must be capable of withstanding

1. Free drop of 9 meters onto an unyielding surface
2. Thermal environment of 800°C for 30 minutes
3. Free drop of 1 meter onto a 15-cm-diameter steel bar without penetration

All packaging must prevent loss of radioactive materials.

Fissile Materials. Fissile materials are the most hazardous materials transported. Safety must be ensured, not only under normal transport, but at the time of an accident. The critical aspect is not only the packaging but also the number of packages that may safely be transported together. The fissile packages are divided into three categories to monitor safe conditions:

1. Fissile Class I—packages that may be transported together in unlimited numbers
2. Fissile Class II—packages that may be transported together only in limited numbers
3. Fissile Class III—packages that must be transported alone

A proper fissile class is either assigned according to Department of Transportation regulations or designated by the Nuclear Regulatory Commission (NRC) or the Department of Energy through approved certificates for the package design.

Specifications known as highway route controlled quantity (HRCQ) have been developed for specific radioactive materials. These materials must be packaged in type B or fissile packaging and are subject to route controls. The HRCQ is defined as a quantity of material that exceeds any of the following:

1. 3,000 times A_1 (special form materials)
2. 3,000 times A_2 (normal form materials)
3. 30,000 Ci

Other Shipping Requirements

A number of other special requirements have been developed for transporting radioactive materials. These include the following:

1. The warning "Radioactive Material" must be displayed on the vessel
2. An identification number must be displayed
3. Name and symbol of each radionuclide must be given
4. A description of the physical and chemical form of the material must be given
5. The activity of each package in Ci, mCi, or uCi, and if the package contains an HRCQ the quantity that falls under HRCQ specifications must be indicated
6. The category A or B must be identified on each package
7. The Transport Index indicating the radioactivity of the material must be given
8. Fissile information must be given
9. Package must have approval identification markings

Each shipping paper must also contain a standardized certification by the shipper that the packages are properly prepared.

Reactor Plant Accidents

Although there have been hundreds of small accidents in the production of electricity using nuclear fuels, there have been only two major accidents. In 1979 the reactor accident at Three Mile Island was controlled before a major meltdown occurred. In contrast, in 1986 the Chernobyl accident in the Soviet Union was a major catastrophe.

Three Mile Island

Dedication ceremonies for the Three Mile Island Unit 2, some ten miles southeast of Harrisburg, Pennsylvania, were held on September 12, 1978. Deputy Secretary of Energy John F. O'Leary, a leading advocate of nuclear power policies, described the new Metropolitan Edison plant as an "aggregation of capital and patience and skill and technology" that was "a miracle in many ways." He added that from this achievement "It is fair to conclude...that nuclear [power] is a bright and shining option for this country" (*Report of the President's Commission on the Accident at Three Mile Island: The Need for Change: The Legacy of*

TMI [New York: Pergamon Press, 1979]). It is thus evident that nuclear power was thought to have a bright future.

The Accident

The Three Mile Island (TMI) accident began at 4:36 A.M. on the morning of March 28, 1979, when a maintenance crew accidentally stopped the flow of water in the main feedwater system of the plant. Within a few seconds the pressure rose, stopping the fission in the reactor. Although fission was now essentially zero, the decaying radioactive materials left from the fission process continued to heat the reactor's cooler water. Because a coolant was not available to control the temperature, it became evident that a loss-of-coolant accident was in progress. As a result, water was leaving the system and the reactor core was on its way to becoming uncovered.

During the next several hours it became evident that a number of serious events were occurring, including an uncontrolled release of radioactivity to the immediate environment. At 1:50 P.M. an explosion of hydrogen gas occurred in the reactor. The significance of the explosion was not recognized until later, and only great good luck prevented a massive explosion.

During the next few days the major effort was simply to cope with the accident. The principal concern was focused on a hydrogen gas bubble that had developed in the reactor and that had the potential of exploding. The bubble gradually disappeared, however, with the gas being distributed through the system. By April 29 the danger of a major explosion was over. The accident at TMI did not end with the lowering of the pressure and heat in the system and the disappearance of the huge gas bubble, nor did the threat to health and safety disappear. Periodic releases of low-level radiation continued. It was several months before the danger of an explosion completely disappeared. Some normalcy gradually returned to the TMI area. The long process to organize and oversee the cleaning up of the TMI-2 accident began—a process that was not complete more than ten years later in 1990, and one that will continue for years to come.

Recommendations of the President's Commission

After the Three Mile Island accident President Jimmy Carter appointed a commission to "make recommendations to enable us to prevent any future nuclear accidents." The commission investigated such aspects as the causes and severity of the accident, handling of the emergency, public and workers' health and safety, the right to information, the role of the Nuclear Regulatory Commission and the role of the electric utilities. (The material quoted in this section is from the *Report of the President's Commission on the Accident at Three Mile Island: The Need for Change: The Legacy of TMI* [New York: Pergamon Press, 1979].)

Commission Findings

The most fundamental findings on the technical aspects of the operation were these:

> The accident at Three Mile Island (TMI) occurred as a result of a series of human, institutional and mechanical failures. Equipment failure initiated the events and contributed to the failure of operational personnel to recognize the actual conditions of the plant. Their training was deficient and left them unprepared for the events that took place. These operating personnel made some improper decisions, took some improper actions, and failed to take some correct actions, causing what should have been a minor incident to develop into the TMI-2 accident.

On the health effects of the accident the commission found:

> On the basis of present scientific knowledge, the radiation doses received by the general population as a result of exposure to the radioactivity released during the accident were so small that there will be no detectable additional cases of cancer, developmental abnormalities, or genetic ill-health as a consequence of the accident at TMI.

The increment of radiation dose to persons living within a 50-mile radius was somewhat less than one percent of the annual background level, which is about 240,000 rems.

The commission report stated further:

> The major health effect of the accident appears to have been on the mental health of the people living in the region of Three Mile Island and of the workers at TMI. There was immediate, short-lived mental distress produced by the accident among certain groups of the general population living within 20 miles of TMI. The highest levels of distress were found among adults: a) living within 5 miles of TMI, or b) with preschool children; and among teenagers: a) living within 5 miles of TMI, b) with preschool siblings, or c) whose families left the area. Workers at TMI experienced more distress than workers at another plant studied for comparison purposes. This distress was higher among the nonsupervisory employees and continued in the months following the accident.

In regard to the protection of public health, the commission found that procedures were lacking or inadequate to handle the problem created by the accident. The Nuclear Regulatory Commission has primary responsibility for health and safety measures as they relate to the operation of commercial nuclear plants. However, the NRC does not require medical examinations of workers other than licensed reactor operators. At the state level a structure to deal with accidents had not been developed. For example, TMI area hospital administrators found no one at the state level with authority to recommend when to evacuate patients and when to assume normal admission practices. Metropolitan Edison was also not prepared for emergencies. For example, during the first days of the accident, Met Ed did not notify its physicians under contract who would have been responsible for the on-site treatment of injured contaminated workers during the accident.

The commission also provided findings on the emergency response. It recognized that planning for the protection of the public in the event of a radiological release was highly complex. It involved the utility and government agencies at the local, state and federal levels. The commission report concluded: "Interaction among NRC, Met Ed, and state and local emergency organizations in the development, review and drill of emergency plans was insufficient to ensure an adequate level of preparedness for a serious radiological incident at TMI."

The commission findings were also critical of Met Ed and its suppliers. In many instances the General Public Utilities Corporation, Met Ed, and B&W (manufacturer of the reactor) failed to acquire enough information about safety problems, failed to analyze adequately what information they did acquire, or failed to act on that information. For example, nine times before the TMI accident, a critical valve stuck open at B&W plants. B&W did not inform its customers of these failures, nor did it highlight them in training programs. In almost all plant operations there were technical deficiencies.

On training operating personnel, the commission report found:

> Training of Met Ed operators and supervisors was inadequate and contributed significantly to the seriousness of the accident. The training program gave insufficient emphasis to principles of reactor safety.
>
> The TMI training program conformed to the NRC standard for training. Moreover, TMI operator license candidates had higher scores than the national average on NRC licensing examinations and operating tests. Nevertheless, the training of the operators proved to be inadequate for responding to the accident.

The president's commission also considered the functions of the Nuclear Regulatory Commission. Its findings were:

> A purpose of the Energy Reorganization Act of 1974 was to divorce the newly created NRC from promotion of nuclear power. According to one of the present NRC commissioners, "I still think it (the NRC) is fundamentally geared to trying to nurture a growing industry." We find that the NRC is so preoccupied with the licensing of plants that it has not given primary consideration to overall safety issues.
>
> NRC labels safety problems that apply to a number of plants as "generic." Once a problem is labeled "generic" the licensing of the individual plant can be completed without resolving the problem.
>
> Although NRC accumulates an enormous amount of information on the operating experience of plants, there was no systematic method of evaluating these experiences and looking for danger signals of possible generic safety problems.

The commission concluded with a report on the public's right to information. It found:

> The quality of information provided to the public in the event of a nuclear plant accident had a significant bearing on the capacity of people to respond to the accident, on their mental health, and on their willingness to accept guidance from responsible public officials.
>
> Before the accident, Met Ed had consistently asserted the overall safety of the plant, although the company had made information concerning the difficulties at TMI-2 public in weekly press releases. This information was not pursued, and often not understood, by the local news media in the area; and the local news media generally failed to publish or broadcast investigative stories on the safety of the plant.
>
> Neither Met Ed nor the NRC had specific plans for providing accident information to the public and news media.
>
> During the accident, official sources of information were often confused or ignorant of the facts. News media coverage often reflected their confusion and ignorance.
>
> Met Ed's handling of information during the first 3 days of the accident resulted in loss of its credibility as an information source with state and local officials, as well as with the news media.

Economic Cost of Three Mile Island

The cleanup of the damaged reactor at Three Mile Island has been more complicated and taken longer than visualized at the time of the accident. Initially projected to take four years and cost $430 million, the cleanup is now targeted for completion in August 1990 at an estimated cost of $973 million.

The cleanup required the development of new equipment, such as tools to pick up debris under 40 feet of water, a remote control arc torch, cameras to peer into dimly lit radioactive spaces, and remote control robots to spray and decontaminate wall surfaces. Technical problems delayed the removal of 700,000 gallons of radioactive coolant water. This water, still in storage and swollen to 2.3 million gallons by additional cleanup work, remains the subject of intense debate. General Public Utilities Corporation (GPU) wants to filter out the remaining cesium, strontium and carbon and boil the water away. Because the water would still contain radioactive tritium, local government officials and environmental activists are concerned.

Besides the direct costs of cleanup at Three Mile Island, the accident placed a high financial burden on GPU, the owner of the facility. GPU lost $172 million on revenues of $1.49 billion the year of the accident. After paying a dividend of $1.20 a share for 1979, the company halted dividends for the next eight years, costing stockholders an estimated $800 million. Its stock price plunged from $18 a share to a low of $3.37. Initially the company had to spend $24 million a month to buy replacement power. The corporation also abandoned a new nuclear facility after spending over $400 million.

After years of financial difficulty the corporation has gained some measure of financial security. Serving 4.2 million customers in New Jersey and Pennsylvania, the efficiency of operations has increased. In 1985 the undamaged nuclear reactor began production of electricity. The company has also negotiated a cost-cutting plan in which it will pay about 10 percent of the cleanup costs and insurance. Its customers, federal and state governments and the nuclear industry will pay the remainder. In 1988 GPU had an after-tax profit of $293.8 million on revenues of $2.83 billion, and its stock price rose above $38.00 a share.

There have been 2,111 cases filed in the courts of Dauphin County. Of these, approximately 300 have been settled, with payments totaling about $25 million. The largest reported settlement was $1.1 million. The remaining cases will continue in the courts for many years.

Health Effects

There is no scientific evidence, according to Dr. George Tokuhata, director of the Division of Epidemiology Research of the Pennsylvania Department of Health, that cancer incidents increased due to the accident. Tokuhata states, "In fact, the rate of leukemia was lower than usual." Concerning studies undertaken by residents, "Those studies don't hold water." Postaccident studies revealed that the most radiation any individual within 50 miles of TMI could have received was about 70 mrems, less than what a person receives each year from natural background radiation. Carol Clawson, spokesman for GPU, adds, "It's difficult to equate emotions with scientific evidence." But, Professor Gordon MacLeod of the University of Pittsburgh, state health commissioner at the time of the accident, says, "It is irresponsible to dismiss possible links between TMI and cancer. We know that some 90 percent of cancers are environmentally induced. We just don't know if there is a connection here." A comprehensive health study of those who lived within five miles of TMI is long overdue.

Robert Reid, who was mayor of Middletown, the site of TMI, at the time of the accident, considers the attitude of the people before the accident.

> I never gave one thought about the accident because we were always told that they had backups to backups and that there never could be an accident. But when the accident took place, the people didn't know how to deal with it. And the people started to use their imaginations, and their imaginations ran away with them. The first thing that most people probably thought was nuclear . . . nuclear bomb . . . the plant will blow up just like an atomic bomb.

The psychological damage and fears remain. Countless pieces of data have been collected about the psychological symptoms of people in the region and about health problems that might be related to radiation. Marjorie and Norman Aamodt, who live 50 miles from the plant, surveyed neighborhoods close to Three Mile Island about symptoms and found cancer deaths up to seven times greater than expected. Mrs. Aamodt states, "General Public Utilities and the Nuclear Regulatory Commission covered up the enormous suffering of people who were exposed to intense plumes of radiation." James Webb, a helicopter pilot who flew television crews over TMI during the accident, has colon cancer. He states, "You can find all kinds of doctors that say radiation doesn't cause my kind of cancer. But you ain't going to find no doctor who's going to get on the stand and raise his hand and say, 'Radiation absolutely did not cause it.'"

Radiation is known to cause genetic effects, including gene mutations and chromosomal abnormalities. To develop definitive answers, much more investigation is needed on a large population sample. A major study has been begun by the Pennsylvania Department of Health. Information has been gathered on 4,000 infants born after March 1979 (11 months after the explosion) to women living within ten miles of Three Mile Island as part of a pregnancy study. Another group of 4,000 infants born between March 1980 and March 1981 have been established as a control group. The analysis covers some 160 variables in each pregnancy, including prenatal care, medical history, and amount of radiation exposure. For this study to be valid, however, the history of these 8,000 babies must be continued throughout their lives.

The fear of health problems will not go away for thousands of central Pennsylvanians. They now see the steam billowing from the mammoth cooling towers of the reactivated, undamaged reactor and wonder when it will occur again. The accident retains a grip on the lives of many in the region. In contrast, while the accident is not forgotten, it does not scare all the people of the region. People have gone on with their lives. Surrounding communities are growing. Economic activity is brisk. TMI has also become a tourist attraction. More than half a million people have gone through the Three Mile Island tourist center and almost 50,000 have toured the plant itself.

Effect on Nuclear Power Development

The TMI accident limited the development of nuclear power in the United States. New nuclear plant orders abruptly ended after the 1979 accident. Since 1979, 116 plants have been canceled and only 109 completed.

The major concern ten years after TMI is safety. In a 1989 survey most people did not approve of nuclear power, and 82 percent would be concerned if they lived near a nuclear plant. A major question that persists is, How safe are nuclear plants?

The nuclear industry has improved its efficiency greatly. As Scott Peters of the nuclear industry's U.S. Council for Energy Awareness states, "We're performing better than we ever have. If you're performing better, you're performing safer." Harold Denton of the Nuclear Regulatory Commission agrees that plants "very definitely are safer today than they were before" the accident.

The confidence of the general public, however, is not reinforced by the continued high number of "incidents" at nuclear plants. Plants must report all "potentially significant safety accidents" ranging from a tool improperly left in an electric cabinet to a flawed reactor. In the four-year period from 1984 to 1987, 11,410 incidents were reported, of which 8,400 were personnel errors. Another measure of efficiency is the number of automatic shutdowns that occur at a plant during a year's operation. In 1984 there were 487 automatic shutdowns, or an average of 5.24 per plant. By 1987 this number had been reduced to 347, or only 3.25 shutdowns per plant. Although the number of automatic shutdowns has been reduced, it still remains quite high. In spite of safety assurances by the nuclear industry, the public remains skeptical, and development remains at a standstill.

The Chernobyl Nuclear Explosion

The Accident

On Saturday, April 25, 1987, at 1:23 A.M., the number 4 reactor of the nuclear power station at Chernobyl exploded, blowing the roof off the plant. A few seconds later a second explosion ejected graphite blocks and fuel material. The hot graphite created a massive fire that threatened to engulf the power plant's three other reactors. After herculean efforts, the fire was contained to the reactor that had exploded.

Distribution of Radioactive Wastes

The radioactive plume of materials and gases—including iodine 131, cesium 137, and strontium 90—was spewed into the stratosphere. The emissions reached a peak on April 26, when about 13 million curies (the amount of any radioactive substance that emits the same number of alpha rays per unit of time as 1 gram of radium) were released from the reactor. By April 27 the radioactive levels had fallen rapidly to about 4 million curies. The amounts declined slowly each day, reaching about 1 million curies on April 30 and May 1. The emission gradually built up to second peak on May 5, when about 7 million curies were emitted. This second peak was caused when a large amount of materials dropping on the core caused it to grow hotter and release more radioactive isotopes. Finally, on the eleventh day after the accident, May 6, liquid nitrogen was placed under the reactor, cooling it. The radiation levels receded to nearly zero by May 7. By this time the reactor fire was under control. But the long job of decontamination and entombing the reactor in concrete had just begun. Years of work will be required to complete the recovery.

Spatial Distribution. The initial evidence outside the Soviet Union that a nuclear accident had occurred came at 2 P.M. on Sunday, April 27, when Chernobyl's radioactive cloud crossed the Swedish border. The next morning, at the Fosmark nuclear power station north of Stockholm, workers undergoing a routine daily radiation examination registered ten times the normal level of radioactivity on their shoe soles. By the afternoon at 4 P.M., Swedish radio reported "10,000 times the normal amount" of cesium 137 in the air.

Within one week after the accident, the nuclear contamination covered an amazingly large area. The spatial pattern was complex due to the shifting wind patterns. During the first day and a half the winds blew northwest, carrying the invisible radioactive cloud over the Soviet republics of Byelorussia, Latvia, and Lithuania, across northern

Poland, and then across the Baltic Sea to the Scandinavian countries. On the second to fourth days the wind shifted to a west and southwesterly flow, blanketing the Ukraine, southern Poland, Austria, Czechoslovakia, southern Germany, Switzerland, northern Italy, and eastern France. The winds also carried the deadly radiation south and west across Romania, Bulgaria, Greece, and Yugoslavia. By the fifth day the winds shifted once again to the northwest, bringing the fallout into central Germany, the Netherlands, and the United Kingdom. Finally, on the sixth day, the winds shifted northeast, carrying the cloud toward Moscow and the central portion of the Soviet Union.

The radioactive material covered at least 20 countries and extended 1,200 to 1,300 miles from the point of the accident. The fallout was extremely uneven. Some parts of Europe directly under the plume received little fallout, and other portions received large amounts. The materials in the plume varied greatly in weight and half-life. The amount of deposition was influenced by rainfall, which washed particles out of the air. Local typography played a role, concentrating radioactive materials and "hot" spots at certain places and dispersing them at others. West German and Swedish scientists found that within a radius of 60 miles, radiation varied by as much as 10 to 15 times.

The Soviets have released only general information on the extent of the fallout. By inference, however, some of the areas affected can be determined. In the vicinity of the plant, the people who were evacuated have not been returned. The towns of Pripyat and Chernobyl have been abandoned and a new town is being built to resettle the displaced population. In the plant's immediate vicinity, clouds were seeded to develop rain and remove as much of the radioactive material as possible. Kiev, the largest city in the vicinity of the accident, with 2.4 million people, received minimal radioactive materials from the plume. The isotopes were driven high into the atmosphere and carried westward away from the city.

Despite aid from the wind currents that carried the plume away from the Ukraine, the people suffered a health and ecological disaster. According to Soviet officials the evacuated people are not likely ever to return home. The soil in the region is the type that will retain radioactive particles for a long time, and the food chain will be contaminated for years.

Effects of the Chernobyl Accident

The effects of the accident must be viewed from two aspects: from within the Soviet Union, and from beyond the Soviet borders.

Effects within the Soviet Union. The effects of the accident were felt immediately at the plant site. Two workers were killed in the initial explosion. By at least April 29, the residents of the nuclear plant city, Pripyat, and three neighboring villages had been evacuated—somewhere between 25,000 and 40,000 people. Altogether about 135,000 people were moved permanently from the area. On May 1, the Soviet government reported that another 18 people were in grave medical condition due to the accident.

On May 8, the Communist party of the Soviet Union (CPSA) published a decree "covering payment and the provision of material benefits to the workers of enterprises and organizations in the zone of the Chernobyl atomic power station." The decree related to the job placement of the evacuees and their rate of pay. The decree further stated the rate of pay for workers involved in the "removal of the consequences of the accident." Two other reports indicated that conditions at the plant were extremely hazardous.

There is evidence that there was much confusion in the Soviet Union on how to handle the accident. The evacuees were frequently treated with callous indifference. *Pravda* observed that the accident had "highlighted bottlenecks" and demonstrated the inability of leaders to act under great stress. In many instances the true nature of the calamity was not revealed. For example, life went on in Kiev as if nothing had occurred at Chernobyl. The Soviet government put the health of 2.4 million people in jeopardy by not making the people aware of the perils of the radioactive emissions.

Besides the immediate damage to the plant, a considerable area of agricultural land and forests was contaminated. One hundred fifty thousand square meters of plastic film were laid on the ground in the vicinity of the plant to preserve vegetation and soil from contamination. Large areas of forest were destroyed and topsoil stripped off to reduce radiation hazards.

The health danger to the vast population under the radioactive emissions of the Chernobyl accident is still not clear. Scientists have estimated that between 3 and 4 percent of the radioactive isotopes in the Chernobyl core were released into the atmosphere. This would be from 50 to 100 million curies of radioactive material. There were at least 50 different types of isotopes with a half-life ranging from a few months to 24,000 years. Cesium is of the greatest concern. Scientists estimate that more cesium entered the environment from Chernobyl than from all the atomic bombs tested in the atmosphere.

A person can be injured by direct exposure to radioactive materials, by inhaling radioactive gases or particles, or by consuming contaminated food or water. All three of these occurred in the

Chernobyl area. As a result of direct exposure there were several hundred immediate deaths. Dr. Robert P. Gale, an American working with Soviet doctors, attempted bone marrow transplants on the most severely irradiated victims. More than half of those treated have died.

The long-term health dangers from fallout are difficult to predict. Scientists vary greatly in their estimates of cancer and resulting deaths. The threat from iodine 131 is difficult to estimate because it disappeared before it could be measured. Even greater uncertainty exists about the impact of low-level radiation at some future date. Estimates now indicate that somewhere between 5,000 and 100,000 fatalities will result from the Chernobyl accident.

Effects beyond Soviet Borders. The response to the increased measurements of radioactivity varied considerably in Europe. Many countries protested the long delay by the Soviets in announcing the magnitude of the accident. The Swedish Nuclear Power Inspector stated, "As a precaution, we would have kept our high-risk groups—children and pregnant women—indoors until the cloud passed." Indoors they would have been shielded from potent but short-lived radioisotopes such as iodine.

By the middle of May more than 20 countries had imposed restrictions on the consumption of fresh vegetables because the levels of radioactivity were above the recommended limits set by health authorities. In addition cattle grazing on contaminated grass were producing milk with radiation above acceptable levels.

Governments were generally not prepared to deal with this emerging problem of contamination. As a result, the restrictions varied greatly. In southern West Germany, the local government of Konstang severely restricted milk and vegetable consumption. In contrast, the adjacent Swiss city of Thugau imposed few restrictions. Sweden developed a national program that discouraged the consumption of fresh vegetables, berries, and freshwater fish. Most of Sweden's dairy cattle were kept in barns until the pastures were tested and declared free of contamination.

In Lapland, in northern Scandinavia, the reindeer pastures were highly polluted due to the heavy rainfall during the passage of the Chernobyl cloud. About 97 percent of the reindeer slaughtered for meat in the late summer had levels of radioactivity above those recommended for human consumption. Norway and Sweden initially banned the use of this meat, but later relaxed the standards somewhat. Because cesium 137 has a half-life of 30 years, the reindeer that graze in the radioactive area will be contaminated for many years. Consumption of reindeer meat could have a cumulative and disastrous health effect. In the eight months following the accident, the normal slaughter would have been

60,000 reindeer. The cost to the local economy was $150 million in that short period. Chernobyl has been an ecological disaster to Lapland, endangering the entire culture of the Sami (Lapp) people.

Most countries were unprepared to cope with the magnitude of the Chernobyl disaster. It was initially reported in France that there was no radioactive fallout, due to air currents moving around the country. Independent observers, however, reported a major rise in the levels of radioactivity. The government finally admitted a mistake in observations with the phony explanation that the officials were on vacation. In Italy, the national government did not act, but left the measurement of radioactivity to local government and citizen groups. In the United Kingdom, the National Radiological Protection Board attempted to minimize possible health effects.

The international agencies did little to provide standards of safety. The International Atomic Energy Agency provided no recommendations on food or health. The World Health Organization provided only a broad warning. The European Economic Community placed restrictions on importing fresh food from Eastern Europe for three weeks, but had no scientific information from which to make judgments.

Emergency Planning and Preparedness for a Nuclear Accident

The general public has a long-standing concern about the safety of nuclear technology. Although society has been quite willing to accept the benefits of nuclear energy and materials, most people would prefer these benefits came risk-free. Because this is not possible, there are two choices. The first is to abandon nuclear technology. The second is to reduce the risk to a minimum and at the same time prepare to deal with an emergency if it materializes. The accidents at Three Mile Island and Chernobyl have heightened the need for effective emergency plans.

An array of projects are under progress in many countries to study the effect of radiation release during a nuclear emergency. New data are continually becoming available, providing the basis for more realistic emergency planning guidelines. In the United States, the Nuclear Regulatory Commission (NRC) has issued a tentative schedule for incorporating research results into reactor licensing.

Current emergency plans for nuclear power plants in the United States consist of two categories of activities:

1. The plans that apply to the plant and its site. These are under the jurisdiction of the private utility and the NRC.
2. The plans that apply to off-site situations. These are under the jurisdiction of local and state authorities, with the Federal Emergency Management Agency (FEMA) coordinating the efforts.

The NRC and the FEMA coordinate all on- and off-site efforts.

The Three Mile Island accident revealed some major deficiencies in emergency planning. It was evident that the plant operators could not identify the start of an abnormal situation in the plant's operation. Further, they were not able to diagnose the problem and implement procedures that would correct the malfunction. The personnel should have been better trained and equipped to evaluate and control a worsening situation. Communications with the media and the general public were also inadequate.

In response to the problems revealed by the Three Mile Island accident, the NRC developed criteria for emergency response. These included:

The need for an on-site technical support center from which reactor operators can obtain immediate technical advice

An operational support center from which on-site emergency response can be coordinated

An emergency operations facility from which the off-site emergency response can be coordinated

A safety parameter display system that operators can use to assess major plant functions to determine plant conditions

This last provision is necessary so that during an accident the plant operator can analyze the total information available in order to make correct decisions. At TMI, the information could not be analyzed quickly, so critical decisions were delayed.

In order to provide information from the total system, Westinghouse has developed a safety-parameter display board based on medical intensive care units, where continuous monitoring of essential body functions is vital. The Westinghouse display system monitors eight important plant parameters, including temperature and pressure. Operators observe the parameters on a display board and can respond rapidly to problems or malfunctions.

The industry has also anticipated future emergencies. The Nuclear Safety Analysis Center collects and analyzes information on equipment malfunctions at all nuclear power plants. In this way the entire industry is made aware of problems and learns how to evaluate individual plans under similar conditions. The Institute for Nuclear Power Operations develops criteria for training nuclear power plant operators.

In recent years there has been much discussion of off-site problems, along with continuing debate as to when a regional evacuation should occur during an accident. Evacuation is very stressful to people and involves real physical dangers. Thus, this is an extremely difficult question. Two conflicting viewpoints prevail. One urges improvements in shielding methods, such as staying indoors and closing all outside air access to houses and buildings, with evacuation occurring only in an extreme emergency. The other viewpoint holds that the population within several miles of the plant should be evacuated as soon as an emergency is declared.

In response to these questions, a hierarchy system has been devised with four classes of emergency action levels, each with its own requirements for on- and off-site responses. This system for emergency response has been developed on the basis of specific plant conditions.

The first class is the least serious and is called an unusual event. It indicates a potential degradation of the plant safety level. An unusual event can normally be handled by plant workers and does not affect public safety. Nevertheless, state and local officials should be informed so that fire and security assistance can be made available if the conditions worsen.

In class two accident conditions have begun to deteriorate so that there is an actual or potential substantial degradation of the plant safety level. In this event, the plant managers would activate the on-site technical support and operational centers and place emergency operation facilities and key on- and off-site emergency personnel on standby status. At this stage public safety is not affected, but local and state officials are notified to begin emergency operations planning in case the situation worsens.

A class three emergency occurs when there is an actual or likely major malfunction of the plant equipment or system. Any release of radioactivity is not expected to exceed Environmental Protection Agency (EPA) protective guidelines except near the site boundary. In a class three emergency, off-site emergency operations would be activated. Local authorities would begin notifying the public of the emergency status and start evacuation planning.

If the safety level further disintegrates and the potential exists for an imminent or actual reactor core melting, then a class four emergency results. Under these conditions there is the potential for loss of containment integrity. Releases of radioactive wastes are likely to exceed EPA guidelines, with both plant and off-site areas being affected. This demands full mobilization of all personnel and facilities, including sheltering the population within five miles of the plant and the possibility of a complete evacuation. If conditions continue to disintegrate, a general public warning is given, and evacuation may be required.

To ensure that an emergency of any class can be handled, the Federal Emergency Management Agency now approves the site-specific plan for local and state governments for each power reactor site to ensure that

1. A formal review of off-site preparedness has been conducted
2. Public meetings have been held for the purpose of reviewing the preparedness status
3. A satisfactory joint exercise has been conducted with both utility and local participation

The plan is reviewed annually.

Health Effects and Exposure Standards for Nuclear Waste

The danger from nuclear waste is very great, so disposal must be absolutely safe. To illustrate, if an individual is within ten miles of an unprotected radioactive waste container from a reactor, he would receive a dose of 500 rems (which is about a 50 percent chance of being fatal) within ten minutes.

To ensure safety, national and international commissions and agencies have set health standards for the measurement of radiation levels. In the United States the relevant organizations are the National Committee on Radiation Protection and Measurements, the Nuclear Energy Commission, and the Environmental Protection Agency.

These standards include the following limitations to exposure. For individuals who are occupationally exposed the maximum possible dose is 5 rems/year of whole body irradiation. If a dose in a single year exceeds this, then in following years the dose must be reduced. The occupational dose to the skin in 1 year must not exceed 15 rems; to the hands, 75 rems; and to the forearms, 30 rems. The maximum permissible exposure for the general public is set at one-tenth the occupational limits, that is, 0.5 rem/year excluding natural background and medical exposures.

Health Effects of Ionizing Radiation

Short-Term Effects

All radiation from radioisotopes has a similar ionizing effect when it touches matter, including biological tissue. In the ionizing process the tissue loses one or more electrons. The ionized atoms absorb a great deal of heat, and in the process chemical changes occur in the biological tissue. These chemical and ultimately structural changes can have serious health consequences.

After extensive research it has been discovered that the health effect of a massive exposure in a short time is quite different from an exposure of about the same amount over a long period of time. For example, if individuals are exposed to 1,000 rems over a short period, immediate severe illness will occur and almost all persons will die within one month. In contrast, if individuals are exposed to ten times the amount of radiation in the natural background over several decades, none will suffer acute illness, and only a few will experience genetic effects causing illness and cancer.

In the use of radiation to treat some diseases, the clinical effects vary with the size of the dose. When the radiation dose on the entire body is below 25 rems, there is no evident chemical effect. In doses of 25 to 100 rems there is a decrease in the number of white blood cells. At doses between 100 and 200 rems there will be a continued reduction in white blood cells, and leukopenia will develop, accompanied by vomiting. At this level the human body will recover in several weeks. If the radiation dose is raised to between 200 and 600 rems, vomiting will occur, accompanied by internal bleeding and complicating infections. Most hair will be lost within two weeks. Treatment will consist of blood transfusions and antibiotics extending over a period of one month to nearly a year. The mortality rate may be as high as 80 percent. When the dose is between 600 and 1,000, rems the treatment is the same, but the mortality rate is between 80 and 100 percent.

Long-Term Effects

The exposure to ionizing radiation may not become evident for many years. The research on these delayed genetic and somatic effects is not as detailed or quantifiable as would be desired. Large human populations have not been exposed to doses of ionizing radiation over extended periods. Although laboratory animals have been intensively studied, the extrapolation from animals to humans is exceedingly difficult. The long-term effects of radiation consist of three categories—genetic effects, human growth and development, and the somatic effects.

The genetic effects of ionizing radiation are most easily recognized in a single, dominant mutation associated with a disease or abnormality. About 1 percent of the population has some form of dominant mutation, such as anemia, dwarfism, or extra fingers or toes. Recessive mutations require that two recessive genes be inherited, one from each parent. These produce such diseases as sickle cell anemia, cystic fibrosis and Tay-Sachs disease. About 0.1 percent of all births are marked by recessive diseases.

It must not be thought, however, that all mutations are caused by doses of ionizing radiation. Most mutations are a response to complex hereditary causes. To measure how many are caused by ionizing radiation, a method is used to estimate the dose that will double the naturally occurring rate of spontaneous mutations. This is referred to as the "doubling dose." Scientific studies indicate that a doubling dose of chronic radiation is between 20 and 200 rems. Because the natural radiation background is only 3 to 5 rems in a reproductive lifetime, only a small percentage of the mutations can be attributed to this effect. Even additional amounts of radiation will have only a small effect on increasing mutations. The possible relationship between additional ionizing radiation and increased disease because of its mutational component is predicted to be between 0.5 and 5 percent in general illness for an additional dose of 5 rems per generation. This appears to be a figure that is much too high. Estimates indicate that the increase in the dose to which the population will be exposed by all nuclear power plants by the year 2,000 will be only 0.001 rem per year or 0.03 rem for a reproductive lifetime.

Evidence has now been accumulated to show that ionizing radiation has a major effect on a fetus in utero and on young children. From studies conducted on subjects of the Nagasaki, Hiroshima, and Marshall Islands nuclear explosions, it was found that there were reduced growth rates, mental retardation, and microcephaly. Different types of doses do produce different health changes. When an individual was exposed within 16 weeks after conception, the individual was more sensitive to health problems than if exposed later. Individuals exposed to 25 rems were mentally retarded, and if the dose was 50 rems, the mental retardation was profound. The individual could not care for himself or carry on a simple conversation. If the radiation occurs after birth, much larger doses are needed over a longer period of time to cause health problems. Evidence is limited, but chronic doses of 365 rems per year or less have not produced health effects.

With the exception of the massive doses of irradiation that produce immediate effects, the somatic effects of small doses over time may not become evident for years or even decades after exposure. The long-term effects may appear as a reduction in fertility, but the most important effect will be the development of cancers such as leukemia or malignancies of the breast, thyroid, stomach, or intestinal tract. Long-term research has now established that the increased risk of cancer is in proportion to the excess in radiation exposure. If a population is exposed to long-term ionizing radiation in excess of the natural background, it is estimated that the total number of additional deaths per million exposed per excess rem would be 10 from thyroid, 50 from breast, 25 to 50 from lung, 20 to 50 from leukemia, and 10 to 15 from gastrointestinal cancers. To place these figures in proper perspective, cancer causes 200,000 deaths for each one million mortality. The cancer mortality due to 10 excess rems in a population of 1 million would be 1,000 in addition to the expected 200,000. (For additional information see Jan A. J. Stolwijk, "Nuclear Waste Management and Risks to Human Health," in Charles A. Walker, Leroy C. Gould, and Edward J. Woodhouse, eds., *Too Hot To Handle? Social and Policy Issues in the Management of Radioactive Wastes* [New Haven, CT: Yale University Press, 1983], pp. 75–93).

2

Chronology

IN THE TWENTIETH CENTURY man has been able to control the energy of the atom. The following critical dates reveal not only the uses of the energy of the atom, but also the environmental hazards that result from these endeavors. For ease of use, five chronologies are presented: Evolution of Atomic Physics; Nuclear Weapons Development; Nuclear Power; Health and Safety Procedures and Standards; and Legislation, Regulations, and Treaties.

Evolution of Atomic Physics

1789 Martin Heinrich Klaproth discovers uranium while experimenting with pitchblende mined in Saxony, Germany.

1841 For decades it was thought that Klaproth had discovered uranium metal. In 1841 Eugene M. Peligot demonstrates that Klaproth's substance is actually the oxide UO_2. This is the beginning of uranium chemistry.

1869 Dmitri I. Mendeleyev in formulating his periodic system identifies uranium as the heaviest of the elements. It holds this position until 1940 and the discovery of the transuranium element.

1879 The modern era of nuclear physics begins when Thomas Crookes achieves the ionization of a gas by an electrical discharge.

1895	William Roentgen discovers penetrating X-rays by use of a discharge tube.
1896	Henri Becquerel discovers that X-rays, now known as gamma rays, can originate from uranium. These rays exhibit the phenomenon of natural radioactivity.
1897	William Thomson identifies the electron as the charged particle responsible for electricity.
c. 1900	Marie Curie and Peter Curie isolate the radioactive element radium.
1905	The revolutionary theory of motion proposed by Albert Einstein provides the foundation for modern nuclear physics. This theory concludes that the mass of an object increases with its speed and is stated by the formula $E = mc^2$, which expresses the equivalence of mass and energy.
1910s	Ernest Rutherford by experimentation finds that the nucleus of the atom is composed of particles carrying a positive charge. He labels these "protons," from the Greek word *protos* meaning "first."
1913	H. G. Wells publishes a futuristic work of fiction, *The World Set Fire.* It is a novel set in the future, 1930, when a new source of energy is obtained through "atomic disintegration." Although this energy could be used for the welfare of mankind, politics intervene and the new energy is used to make weapons of destruction.
1932	James Chadwich discovers that one of the components of the nucleus of the atom is an infinitesimally small, electrically chargeless bit of matter, which he calls a "neutron."
	Les Szilard begins experimentation to find an element that can be split, releasing energy—and neutrons—that will then go on to split other atoms, which in turn will release other energy and more neutrons—and so on. This is a chain reaction.

1938 Otto Hahn and Fritz Strassman bombard a stream of neutrons at a uranium metal. At the conclusion of the experiment it is found that mixed with the uranium target is a trace amount of barium, a light element. Lisa Meitner, a nuclear physicist, explains that the phenomenon is due to the "splitting" of the heavy uranium atom, forming a number of lighter elements.

1939 Edward Teller and Niels Bohr confirm the experiment of Otto Hahn in a laboratory at the Carnegie Institute's Department of Terrestrial Magnetism.

Researchers at the University of Minnesota manage to isolate a new isotope of uranium, U-235. It is lighter than the uranium found in nature, U-238. It is also more unstable and more fissionable, and experiments indicate it will be a perfect weapons material.

1940 Scientists in Berkeley, California, under the direction of Glenn Seaborg isolate a man-made element, plutonium, that is an even better material for fission.

1941 December 6. President Franklin Roosevelt appoints the S-1 Committee to study the feasibility of an atomic weapon.

1942 The S-1 Committee reports that an atomic bomb is feasible. The cost would be $100 million and the bomb could be built by July 1944.

Brigadier General Leslie Groves assumes control of the Manhattan Project, a huge facility at Oak Ridge, Tennessee, built for the purpose of separating U-235 from U-238. Also in the project is a reactor in eastern Washington to take U-235 and from it produce large quantities of the more fissionable plutonium.

J. Robert Oppenheimer, physicist from Berkeley, is given the responsibility to assemble the atomic bomb team at the Manhattan Project.

1945 July 16. The first atomic bomb is detonated at 5:29:45 near Alamogordo, New Mexico.

Nuclear Weapons Development

1945 August 6. First atomic bomb, "Little Boy," is dropped on Hiroshima, Japan.

1945 *cont.* August 9. Atomic bomb "Fat Man" is dropped on Nagasaki, Japan.

1946 July 24. United States conducts the world's first underwater nuclear weapons test. The weapon is detonated at Bikini Atoll in the Pacific Ocean with a yield of 23 kilotons.

1948 June 1. U.S. Navy tests the P2V Neptune aircraft to determine its suitability to deliver nuclear weapons from aircraft carriers.

1950 January 31. President Truman announces program to develop the thermonuclear bomb.

1951 October 3. United Kingdom detonates its first nuclear weapon at Monte Bello in Australia as part of Operation Hurricane.

1952 October 31. United States detonates its first thermonuclear device on the surface of Eniwetok Atoll in the South Pacific. The test, named *Mike*, is the largest-yield nuclear weapon tested to date at 10.4 megatons.

1954 February 28. The United States detonates a bomb with a yield of 15 megatons at Eniwetok Atoll. The "experimental thermonuclear device" exceeds its predicted yield and, with a shift of wind direction, carries lethal radioactive fallout to a large ocean area, where a Japanese fishing boat is contaminated.

Atomic Energy Commission (AEC) detonates six test weapons in the Marshall Islands.

March 26. United States detonates Shot Romeo at Operation Castle on a barge in Bikini Atoll. The test, with a yield of 11 megatons, is the first of 36 weapons tests in barges moved to Bikini.

November 30. The USS *Nautilus*, the first nuclear-powered submarine, is commissioned.

1955 February 15. AEC releases report "Effects of High-Yield Nuclear Explosions."

1956 First U.S. nuclear-powered submarine, *Nautilus,* has radiological accident due to failure of shielding.

1957 May 15. United Kingdom detonates its first thermonuclear weapon as part of Operation Grapple.

December. Soviet Union launches the first nuclear-powered surface ship, icebreaker *Lenin,* with three nuclear reactors as the power source.

1958 August 22. President Eisenhower places a moratorium on U.S. nuclear weapons tests.

December 5. U.S. weapons system test ship USS *Observation Island* is commissioned. This is the first ship capable of supporting full flight and system tests for the submarine-launched ballistic missile (SLBM) development program.

1959 April 20. First successful U.S. Polaris missile flight test.

August 27. First seaborne launch of a solid-propellant ballistic missile, from USS *Observation Island* off the coast of Florida.

1960 February 13. France tests its first nuclear weapon at Reggan in the Algerian Sahara Desert.

August 20. First launch of Polaris test missile from a submerged submarine, the USS *George Washington,* off Cape Canaveral, Florida.

October 10. First test of Polaris A-2 missile, successfully launched at Cape Canaveral, Florida, traveling over 1,400 nautical miles.

1962 August 7. First flight test of U.S. Polaris A-3 SLBM at Cape Canaveral.

1963 First deployment of Soviet Kashin class guided-missile destroyer (DDG), which can carry nuclear-capable 21-inch torpedoes.

April 6. United States and United Kingdom sign the Polaris Sales Agreement establishing the terms for selling U.S. SLBM equipment to Britain.

1964 China builds its only known Golf class submarine at Darien, Manchuria.

October 16. China detonates its first nuclear weapon, an implosion fission bomb of about 20-kiloton yield.

1966 July 2. France begins the first of its nuclear weapons tests at the Pacific Test Center of Mururoa and Fangataufa Atolls.

China launches first CSS-1 medium-range ballistic missile (MRBM) from Shuang Ch'eng-tze launch center in Kansu Province.

1967 March 29. France launches its first ballistic missile submarine, the nuclear powered *Le Redoutable.*

June 17. China tests its first thermonuclear weapon, a fusion bomb of 3-megaton yield, at Lop Nor in Sinkiang Province.

July. First Soviet Moskva class helicopter carrier is deployed.

1968 August 24. France detonates its first thermonuclear weapon at Fangataufa Atoll in French Polynesia in the Pacific.

Nuclear Power

1948 August 5. Atomic Energy Commission (AEC) establishes a Division of Reactor Development.

1950s AEC provides generous incentives for discovery and development of uranium deposits to have adequate fuel for civilian and military purposes.

1951 December 20. First generation of electric power from a nuclear reactor at AEC's reactor testing center in Arco, Idaho.

1953 December 8. President Dwight Eisenhower delivers "Atoms for Peace" speech before the United Nations.

1954 January 24. U.S. Navy launches first atomic submarine, U.S.S. *Nautilus.*

1954 *cont.* March 14. AEC and Duquesne Light Company negotiate an agreement to construct jointly a pressurized water reactor demonstration facility at Shippingport, Pennsylvania.

1955 March 30. AEC creates Division of Licensing for nuclear energy.

1955–1957 AEC issues several new regulations dealing with nuclear power under Title 10 of the Code of Federal Regulations.

1956 January 6. Power Reactor Development Company (PROC) applies to AEC for a construction permit to build a commercial fast-breeder reactor in Lagoona Beach, Michigan.

AEC issues first construction permits for two large-scale nuclear power facilities by Consolidated Edison at Indian Point, New York, and Commonwealth Edison at the Kankakee and Des Plaines river junction, southwest of Chicago, Illinois.

1957 The world's first generating nuclear power plant began operating at Shippingport, Pennsylvania.

1973 French government announces plans to produce 70 percent of the nation's electricity and 30 percent of the total energy consumption of the nation by nuclear power by 1990.

1979 March 28. Three Mile Island accident at Middletown, Pennsylvania.

October 30. President's commission presents findings on the Three Mile Island accident.

1987 April 25. Chernobyl nuclear explosion.

1989 The world's first nuclear power plant at Shippingport, Pennsylvania, is dismantled.

Health and Safety Procedures and Standards

1895 October 18. Modern radiology begins when Karl Wilhelm Roentgen notices unknown greenish fluorescence on his workbench while testing a Crooker tube. The light was produced by unknown rays from the tube striking a screen treated with barium platinocyanide.

1896 January 23. Roentgen gives the first public lecture on X-rays at Wurzburg, Germany, demonstrating how the rays permitted photographs to be made of the bones in a colleague's hand.

1920–1930 X-ray therapy increases in importance because of the relatively low power of radium implants and the capacity of X-rays to multiply its force.

1936 Isotopes first used to study metabolism and to treat leukemia. They prove especially valuable for studying gland functions, tumors' locations, and blood conditions. After 1945, isotopes will be produced in quantity by atomic fission.

1947 June. Reactor Safeguard Committee established to advise the Atomic Energy Commission (AEC) on the hazards of reactor operation.

1953 March 20. National Committee on Radiation Protection (NCRP) publishes recommendations on maximum permissible amounts of radioisotopes in the human body and maximum permissible concentrations in air and water.

August 9. AEC merges the Reactor Safeguard Committee and the Industrial Committee on Reactor Location Problems into the Advisory Committee on Reactor Safeguards (Safeguards Committee).

1954 September 24. NCRP publishes recommendations on maximum permissible doses of radiation from external sources.

1955 February 15. AEC releases report on "Effects of High-Yield Nuclear Explosions."

1955 *cont.* February 22. Seven-state Uranium Mining Conference on Health Hazards establishes "working levels" for radon and radon daughters in uranium mines.

November 6. Reactor Hazards Evaluation Staff transferred to the Division of Civilian Application.

1956 June 12. National Academy of Sciences issues report on "The Biological Effects of Atomic Radiation."

1957 May 27–June 7. Congressional Joint Committee holds hearings on the "Nature of Radioactive Fallout and Its Effects on Man."

September 2. Price-Anderson amendment to 1954 Atomic Energy Act becomes law. Provides government indemnity in event of major reactor accident. Safeguards Committee made a statutory body.

1959 March 26. Surgeon general's National Advisory Committee on Radiation recommends that primary authority for radiation safety be vested in the U.S. Public Health Service.

August 14. President Eisenhower creates the Federal Radiation Council to provide guidance to agencies on radiological health protection.

September 12. Office of Health and Safety established.

1960 May 13. Federal Radiation Council submits radiation-protection guidelines. Approved by President Eisenhower.

December 15–16. Federal officials and governors of uranium-mining states meet to discuss radiation hazards in mines.

1961 September 20. President Kennedy approves Federal Radiation Council's proposal to change guidelines for exposure to strontium 89, strontium 90, iodine 131, and radium 226. AEC subsequently modifies its regulations.

1971 U.S. Congress enacts a National Cancer Act. The act includes a research program, the creation of a new cancer institute independent of the National Institute of Health, a provision for clinical application of research findings, an international cancer data bank, and a cancer panel that reports directly to the president.

1981 Japan's study of the A-bomb and environmental hazards of radiation published in the United States.

1988 September 23. The Environmental Protection Agency issues a national health advisory on radon recommending testing of homesites for the presence of the cancer-causing gas.

Legislation, Regulations, and Treaties

1945 Declaration of the President of the United States, the Prime Minister of the United Kingdom, and the Prime Minister of Canada. An initial statement to control nuclear energy so that it will be developed for peaceful purposes rather than for military uses.

Moscow Conference of the Three Foreign Ministers of the United States, United Kingdom, and the Soviet Union. The foreign ministers propose a Commission for the Control of Atomic Energy.

1946 Atomic Energy Act of 1946 becomes law. Establishes the Atomic Energy Commission (AEC). Transfers Manhattan Project's programs and facilities to the five-member civilian commission.

1953 Atoms for Peace. President Eisenhower announces his plans for the peaceful use of the atom. One of the first endeavors in the Cold War to neutralize the efforts of the Soviets.

1954 Atomic Energy Act of 1954. AEC given expanded regulatory responsibilities.

1959 Antarctic Treaty signed, agreeing that the Antarctic will be used only for peaceful purposes.

1962 Treaty of Tlatelolco. The treaty establishes the policy of nuclear-weapons-free zones in the world.

1963 Limited Test Ban Treaty signed by the United States, United Kingdom, and the Soviet Union. Bans nuclear tests in the oceans, in the atmosphere, and in outer space.

1967 Treaty for the Prohibition of Nuclear Weapons in Latin America (Tlatelolco Treaty) signed. Treaty went to nations for ratification on 22 April 1968. This is the first arms control treaty creating a nuclear-weapons-free zone.

1968 Treaty on the Non-Proliferation of Nuclear Weapons. An international treaty to control the proliferation of nuclear weapons.

1971 Sea-bed Treaty signed, preventing the emplacement of nuclear weapons on the ocean seabed. Entered into force on 18 May 1972.

1972 Uranium Radiation Exposure Remedial Action. Congress authorizes an appropriation to provide financial assistance to undertake remedial action to limit the exposure of individuals to radiation emanating from uranium mill tailings in the Grand Junction, Colorado, area.

May 25. Signature and entry into force of the United States–Soviet Union Agreement on the prevention of incidents on and over the oceans.

1974 Hazardous Materials Transportation Act. Establishes criteria for handling hazardous materials.

Energy Reorganization Act of 1974.

1978 Nuclear Non-Proliferation Act. Indicates that nuclear explosives pose a grave threat to the security of the United States.

Uranium Mill Tailings Radiation Control Act of 1978. Authorizes the secretary of energy to enter into cooperative agreements with certain states respecting residual radioactive material at existing sites and to provide for the regulation of uranium mill tailings under the Atomic Energy Act of 1954.

1980 Low-Level Radioactive Waste Policy Act. Defines low-level waste as radioactive waste not classified as high-level. A major provision places the responsibility for disposal of low-level radioactive waste on the states.

October 1. West Valley Demonstration Project Act. Authorizes the Department of Energy to carry out a project at the Western New York Service Center in West Valley, New York, demonstrating how to manage high-level liquid nuclear waste.

1981 Consumer-Patient Radiation Health and Safety Act. Provides for the establishment of minimum standards by the federal government for the accreditation of education programs for persons who administer radiologic procedures and for the certification of such persons.

1983 January 7. The Nuclear Waste Policy Act of 1982. Provides for the disposal of high-level radioactive waste and spent nuclear fuel, for a program of research, development, and demonstration regarding the disposal of high-level radioactive waste and spent nuclear fuel, and for other purposes.

1985 Low-Level Radioactive Waste Policy Amendments Act. Amends 1980 act to improve procedures for implementation of compacts providing for the establishment and operation of regional disposal facilities for low-level radioactive waste and to grant the consent of the Congress to certain interstate compacts on low-level radioactive waste.

Federal Nuclear Waste Disposal Liability Act. Establishes liability and indemnification for nuclear incidents arising out of federal storage, disposal, and related transportation of high-level radioactive waste and spent nuclear fuel.

August 5. South Pacific Nuclear-Free Zone Treaty. The preamble expresses the commitment to world peace, with every country having an obligation to strive for the elimination of nuclear weapons, a belief in the efficacy of regional arms control measures, and a reaffirmation for halting nuclear proliferation.

1987 Nuclear Protection and Safety Act. Develops policies for technical operations of facilities operated by the Department of Energy.

3

Laws, Regulations, and Treaties

BECAUSE THE DEVELOPMENT OF ATOMIC ENERGY has both national and international implications, national laws and international treaties have been developed. The basic philosophy behind all of these endeavors is to channel atomic energy into peaceful endeavors for the benefit of the world's peoples and to control the use of nuclear weapons. This chapter considers the major acts of the United States Congress in providing policies for the development of atomic energy and the international agreements and treaties to limit the use and spread of nuclear weapons and to encourage the use of atomic energy for peaceful endeavors.

Atomic Energy Act

The Atomic Energy Act established a national policy for the development and regulation of the atomic industry for both nuclear weapons and peaceful operations and made the Atomic Energy Commission responsible for both functions. A key section of the law encouraged "widespread participation in the development and utilization of atomic energy for peaceful purposes and to the maximum extent consistent with the common defense and security and with the health and safety of the public."

Atomic Energy Act of 1946 and as Amended in 1954 and 1980

The initial act indicated that

> Research and experimentation in the field of nuclear chain reactions have attained the stage at which the release of atomic energy on a large scale is practical. The significance of the atomic bomb for military purposes is evident. The effect of the use of atomic energy for civilian purposes upon the social, economic and political structure of today cannot now be determined. It is a field in which unknown factors are involved. . . . It is reasonable to anticipate, however, that tapping this new source of energy will cause profound changes in our present way of life. Accordingly, it is hereby declared to be the policy of the people of the United States that, subject at all times to the paramount objective of assuring the common defense and security, the development and utilization of atomic energy shall, so far as practicable, be directed toward the standard of living, strengthening free competition in private enterprise, and improving world peace.

The act established five major programs to develop atomic energy. These were

1. A program of assisting and fostering private research and development to encourage maximum scientific progress
2. A program for the control of scientific and technical information that would permit the dissemination of such information to encourage scientific progress and for the sharing on a reciprocal basis of information concerning the practical industrial application of atomic energy as soon as effective and enforceable safeguards against its use for destructive purposes could be devised
3. A program of federally conducted research and development to assure the government of adequate scientific and technical accomplishments
4. A program for government control of the production, ownership, and use of fissionable material to ensure the common defense and security and to ensure the broadest possible exploitation of the fields
5. A program of administration that would be consistent with the foregoing policies and with international arrangements made by the United States, and that would enable the Congress to be currently informed so as to take further legislative action as might thereafter be appropriate.

The act established the Atomic Energy Commission, to be composed of five members. The commission was appointed by the president with the advice and consent of the Senate. There was also a General Advisory Committee of nine members from civilian life to provide advice relating to materials, production, and research and development. A Military Liaison Committee from the Department of War and Navy was also formed.

The act established policies and regulations for the development and utilization of atomic energy.

Research

The commission was authorized to develop a program of research relating to

1. Nuclear processes
2. The theory and production of atomic energy, including processes, materials, and devices related to such production
3. Utilization of fissionable and radioactive materials for medical, biological, health, and military purposes
4. Utilization of fissionable and radioactive materials and processes entailed in the production of such materials for all other purposes, including industrial uses
5. The protection of health during research and production activities

Production of Fissionable Material

The act established the basis for the production of fissionable materials in the United States. Specifically, it stated:

> It shall be unlawful for any person to own any facilities for the production of fissionable material or for any person to produce fissionable material.

The act further indicated that the commission was to be the exclusive owner of all facilities for the production of fissionable material, except for research and development activities. At no time was a private producer of fissionable material to produce a sufficient quantity to produce an atomic bomb.

Control of Materials

The act established the policy that the Atomic Energy Commission was the sole owner of all fissionable material. The commission was authorized to distribute fissionable material owned by it, with or without charge, to applicants requesting fissionable material. The commission could also purchase fissionable material from any source in the world.

Military Application of Atomic Energy

The Atomic Energy Commission was authorized to

1. Conduct experiments and do research and development work in the military application of atomic energy
2. Engage in the production of atomic bombs, atomic bomb parts, or other military weapons utilizing fissionable materials; except that such activities were to be carried on only to the extent that the express consent and direction of the president of the United States had been obtained, which consent and direction should be obtained at least once each year.

Utilization of Atomic Energy

The 1946 act placed the utilization of atomic energy under control of the commission. When atomic energy has been sufficiently developed, the commission will report to the president stating all the facts with respect to such use, including an estimate of the social, political, economic, and international effects of atomic energy.

International Arrangements

The act recognized the need to coordinate with all international agreements.

Control of Information

The act established an absolute control of the dissemination of information. The act stated:

> Until Congress declares that effective and enforceable international safeguards against the use of atomic energy for destructive purposes have been established, there is to be no exchange of Restricted Data with other nations with respect to the use of atomic energy. The dissemination of scientific and technical information relating to atomic energy shall be permitted.

Disposal of Radioactive Nuclear Waste

The Congress of the United States has passed three fundamental laws providing regulations on the disposal of radioactive nuclear wastes. The first established a demonstration project at West Valley, New York. Congress recognized that there are two basic types of radioactive wastes—low level and high level. The Nuclear Waste Policy Act of 1982 was directed to solving the problem of high-level waste by the creation of safe high-level-waste depositories. The Low-Level Radioactive Waste Policy Act of 1980 established a procedure for developing low-level radioactive waste sites. In 1985, through the Federal Nuclear Waste Disposal Liability Act, the federal government assumed responsibility for liability and indemnity for certain types of nuclear accidents.

West Valley Demonstration Project Act

In 1980 Congress passed the West Valley Demonstration Project Act, which authorized the Department of Energy to carry out a project at its Western New York Service Center in West Valley, New York, demonstrating how to manage high-level liquid nuclear waste.

The project called for performing the following activities:

1. Solidify, in a form suitable for transportation and disposal, high-level radioactive waste by vitrification or by such other technology determined to be most effective.
2. Develop containers suitable for the permanent disposal of solidified high-level radioactive waste.
3. Transport, as soon as feasible, and within the provisions of the law, the solidified waste to an appropriate federal repository for permanent disposal.
4. In accordance with applicable licensing requirements, disposal of low-level radioactive waste and transuranic waste produced by the solidification of the high-level radioactive waste under the project.
5. The storage containers of the solidified high-level radioactive wastes will be decontaminated and decommissioned.

Nuclear Waste Policy Act of 1982

The Nuclear Waste Policy Act of 1982 is based on findings by Congress that

1. Radioactive waste creates potential risks and requires safe and environmentally acceptable methods of disposal.
2. A national problem has been created by the accumulation of (A) spent nuclear fuel from nuclear reactors; and (B) radioactive wastes from (1) reprocessing of spent nuclear fuel; (2) activities related to medical research, diagnosis, and treatment; and (3) other sources.
3. Federal efforts during the past 30 years to devise a permanent solution to the problems of civilian radioactive waste disposal have not been adequate.
4. Although the federal government has the responsibility to dispose of nuclear waste to protect the public health, safety and environment, the cost of the disposal is the responsibility of the generators and owners of the waste and spent fuel.
5. The generators of the waste must also pay for the interim storage.
6. There must be state and public participation in the planning and development of repositories in order to promote public confidence in the safety of the disposal processes.

In order to solve the disposal problem, the purposes of the 1982 act were

1. To establish a schedule for the siting, construction, and operation of repositories that would provide a reasonable assurance that the public and the environment would be protected from the hazards posed by high-level radioactive waste
2. To establish federal responsibility and a policy for disposal of waste and spent fuel
3. To define the relationship between the federal and state governments
4. To establish a nuclear-waste-fuel fund made up of payments by the generators and owners of the waste.

Site Characteristics

The act provides guidelines for the selection of the disposal sites. Of the considerations, geologic conditions are to be of primary importance. The guidelines must specify factors that qualify or disqualify any site, including the location of valuable national resources, hydrology, geophysics, seismic activity and atomic energy defense activities, proximity to water supplies, population, and national parks. Any site will be disqualified if it is located in a highly populated area or adjacent to an area one mile by one mile having a population of not less than 1,000 persons. The site must also consider transportation of the waste or spent nuclear fuel to it. The act stipulated that the secretary of the Department of Energy would nominate a site not later than July 1, 1989. This goal was not achieved.

Participation of States

The act requires that the secretary of energy notify a state of the selection of a site within its borders before detailed geologic data gathering begins. If a state does not approve, the governor or legislature may submit a notice of disapproval to Congress within 60 days after the date that the president recommends such a site to Congress.

If the state approves, the secretary of energy will provide financial and technical assistance in evaluating the site of the repository. Further, the secretary must notify the state prior to the transportation of any high-level radioactive waste and spent nuclear fuel into the state for disposal in the depository.

Interim Storage Program

Congress recognized that high-level waste and spent nuclear fuel had accumulated over the years. In order to deal with this problem Congress indicated that

1. The persons owning and operating civilian nuclear power reactors have the primary responsibility for providing interim storage of spent nuclear fuel from such reactors, by maximizing, to the extent practical, the effective use of existing storage facilities of the site of each civilian nuclear power reactor, and by adding new on-site storage capacity in a timely manner when practical.
2. The federal government has the responsibility to encourage and expedite the effective use of existing storage facilities and the addition of needed new storage capacity at the site of each civilian nuclear power reactor
3. The federal government has the responsibility to provide not more than 1,900 metric tons of capacity of interim storage of spent nuclear fuel for civilian nuclear power reactors that cannot reasonably provide adequate storage capacity at the sites of such reactors when needed to ensure the continued, orderly operation of such reactors.

In solving the problem of interim storage, the purposes of the act were:

1. To provide for the utilization of available spent-nuclear-fuel pools at the site of each civilian nuclear power reactor to the extent practical and the addition of new spent-nuclear-fuel capacity when practical at the site of the reactor
2. To provide for the establishment of a federally owned and operated system for the interim storage of spent nuclear fuel at one or more facilities owned by the federal government, with not more than 1,900 metric tons of capacity to prevent disruption in the orderly operation of any civilian nuclear power reactor that could not reasonably provide adequate spent-nuclear-fuel storage capacity at the site of such reactors when needed.

Monitored Retrievable Storage

Congress recognized that a permanent solution to the storage of high-level radioactive waste and spent nuclear fuel could extend over a long period. It found that

1. Long-term storage of high-level radioactive waste or spent nuclear fuel in monitored retrievable storage facilities is an option for providing safe and reliable management of such waste or spent fuel.
2. The executive board and Congress should proceed as expeditiously as possible to consider fully a proposal for construction of one or more monitored retrievable storage facilities to provide such long-term storage.
3. The federal government has the responsibility to ensure that site-specific design for such facilities is available.
4. The generators and owners of the high-level waste and spent nuclear fuel to be stored in the facilities must pay the cost of the long-term storage.
5. Permanent disposal facilities should not be delayed by monitored retrievable storage facilities.

Low-Level Radioactive Waste Policy Act of 1980

The act defined low-level radioactive waste as radioactive waste not classified as high-level radioactive waste, transuranic waste, spent nuclear fuel, or certain by-product material. The policy of the federal government in the disposal of low-level radioactive wastes, as established by this act, is:

1. Each state is responsible for providing for the availability of capacity either within or outside the state for the disposal of low-level radioactive waste generated within its borders except for wastes generated as a result of defense activities of the secretary or federal research and development activities.
2. Low-level radioactive waste can be most safely and efficiently managed on a regional basis.

In order to carry out this policy the states may enter into such compacts as may be necessary to provide for the establishment and operation of regional disposal facilities. The compacts must be approved by Congress and reviewed every five years, when Congress may withdraw its consent. Further, after 1 January 1986 the compact may restrict the use of the regional facility to waste generated within the region.

The secretary of energy under the act had to define the disposal capacity needed for present and future low-level waste sites. He was also required to prepare an evaluation of each license site, including operating history, an analysis of the adequacy of disposal technology employed at each site, and recommendations to ensure public health and safety from wastes transported to such sites. The secretary also had to evaluate on a regional basis the transportation requirements for safely moving the low-level radioactive wastes.

Low-Level Radioactive Waste Policy Amendments Act of 1985

These amendments were enacted to improve procedures for the implementation of compacts providing for the establishment and operation of regional disposal facilities for low-level radioactive waste and to give the consent of Congress for establishing certain interstate compacts on low-level radioactive wastes. The federal government recognized that it was responsible for the disposal of

1. Low-level radioactive waste owned or generated by the Department of Energy
2. Low-level radioactive waste owned or generated by the United States Navy
3. Low-level radioactive waste owned or generated by the federal government as a result of any research, development, testing, or production of an atomic weapon

4. Any other low-level radioactive waste with concentrations of radionuclides that exceed the limits established by the commission

With the passage of the act the secretary of energy had to submit to Congress a comprehensive report setting forth the recommendations for the safe disposal of low-level radioactive waste. The report had to include:

1. An identification of the radioactive waste involved, including the source of the waste, volume, concentration, and other relevant characteristics
2. An identification of federal and nonfederal options for disposal
3. A description of the action proposed, including projected costs
4. An identification of any statutory authority required for the disposal of such waste

The act indicated that it was the policy of the federal government that the responsibilities of the states for disposal of low-level radioactive wastes could be most safely and effectively managed on a regional basis. In order to accomplish this, states could enter into interstate compacts.

Title II of the act known as the Omnibus Low-Level Radioactive Waste Interstate Compact Consent Act provided the procedures by which Congress gave consent for the establishment of regional compacts. The regional compacts established as of 1988 were:

1. Northwest Interstate Compact on Low-Level Radioactive Waste Management. States in the compact were Alaska, Hawaii, Idaho, Montana, Oregon, Utah, Washington, and Wyoming.
2. Central Interstate Low-Level Radioactive Waste Compact. States included: Arkansas, Iowa, Kansas, Louisiana, Minnesota, Missouri, Nebraska, North Dakota, and Oklahoma.
3. Southeast Interstate Low-Level Radioactive Waste Management Compact. States included: Alabama, Florida, Georgia, Mississippi, North Carolina, South Carolina, Tennessee, and Virginia.
4. Central Midwest Interstate Low-Level Radioactive Waste Compact. States included: Illinois and Kentucky.

5. Midwest Interstate Low-Level Radioactive Waste Management Compact. States included: Iowa, Indiana, Michigan, Minnesota, Missouri, Ohio, and Wisconsin.
6. Rocky Mountain Low-Level Radioactive Waste Compact. States included: Arizona, Colorado, Nevada, New Mexico, Utah, and Wyoming.
7. Northeast Interstate Low-Level Radioactive Waste Management Compact. States included: Connecticut, New Jersey, Delaware, and Maryland.

The policies and regulations for each of the interstate compacts are similar. As a policy it is recognized that low-level radioactive wastes are generated by essential activities and services that benefit the citizens of the states. It is further recognized that the protection of health and safety of the citizens of the compact states and the most economical management of low-level radioactive wastes can be accomplished by minimizing the amount of handling and transportation required to dispose of the wastes through the cooperation of the states in providing facilities that serve the region.

Each of the states participating in the compact has agreed to permit transportation of low-level radioactive wastes in the region. These practices include:

1. Maintaining an inventory of all generators of low-level wastes within the state
2. Conducting unannounced inspections of low-level waste generators and carriers
3. Authorizing containers for shipment of wastes
4. Setting penalities for violations of parking and transportation standards

A low-level radioactive waste facility will accept waste from any state within the compact region. However, no facility can accept waste from outside its designated region. Each of the compact regions is required to establish a committee or commission to oversee the operation of the low-level waste facility.

The Federal Nuclear Waste Disposal Liability Act of 1985

The Federal Nuclear Waste Disposal Liability Act of 1985 amends the Price-Anderson provisions of the Atomic Energy Act of 1954. It establishes liability and indemnification for nuclear incidents arising out of federal storage, disposal, or related transportation of high-level radioactive waste and spent nuclear fuel.

The act is based on two congressional findings:

1. The federal government currently does not, but should, assume the responsibility to provide total indemnification for public liability claims arising out of nuclear incidents relating to federal storage, disposal, and related transportation of high-level radioactive waste and spent nuclear fuel.
2. The indemnification should be paid through the Nuclear Waste Fund.

The aggregate amount of payments is not to exceed $5 billion for each nuclear incident.

Transportation of Nuclear Material

Within the nuclear fuel cycle, nuclear materials are transported throughout the processes, beginning with the hauling of uranium ores to mills and ending with the disposal of spent nuclear fuels. Congress has recognized that this type of transportation has unique conditions that can affect public health and welfare. Congress addresses these problems in the Hazardous Materials Transportation Act of 1974.

Hazardous Materials Transportation Act of 1974 (Amended)

The Transportation Safety Act of 1974 has a number of titles, of which Title I is Hazardous Materials. Hazardous materials includes a wide category of radioactive materials. This act established criteria for handling hazardous materials. The secretary of transportation is authorized to develop a set of standards to include:

1. Minimum number of personnel needed to transport nuclear wastes
2. Minimum level of training and qualification of personnel
3. Type and frequency of inspection of transport vessels
4. Equipment to be used for detection, warning, and control of risks of such materials
5. Specifications regarding the use of equipment and facilities used in handling and transporting such materials

6. A system for monitoring safety assurance procedures for the transportation of such materials.

Radiation from Mill Tailings

Congress has recognized the problem of radiation from mill tailings by enactment of a number of pieces of legislation. In 1972 a bill was enacted to remove uranium tailings from specific sites at Grand Junction, Colorado. By 1978 it was recognized that a broader problem existed and it had to be treated from a national viewpoint. This culminated in the Uranium Mill Tailings Radiation Control Act of 1978.

Uranium Radiation Exposure Remedial Action

In June 1972 Congress authorized an appropriation to the Atomic Energy Commission in accordance with the Atomic Energy Act to provide financial assistance to undertake remedial action to limit the exposure of individuals to radiation emanating from uranium mill tailings that had been used as construction material in the area of Grand Junction, Colorado. The secretary of energy was authorized to enter into a cooperative agreement with the state of Colorado. The authorization required that:

1. Remedial guidelines be published by the surgeon general of the United States
2. Property owners apply within eight years of the enactment of the act, and that the remedial activity be performed by the state of Colorado or its authorized contractors and be paid by the state
3. The United States be released from any mill tailings liability or claims upon completion of the remedial action
4. The state of Colorado retain custody of and responsibility for any uranium mill tailings removed from the site

The act authorized $5,000,000 to complete the removal of the mill tailings. The federal government was to provide up to 75 percent of the cost of the state's program.

Uranium Mill Tailings Radiation Control Act of 1978

After radioactive materials had been mined for years in the United States, it was found that uranium mill tailings located at active and inactive mill sites could pose a potential and significant radiation health hazard. To protect the public health and welfare, Congress recognized that the mill tailings had to be stabilized and controlled to prevent or minimize radon diffusion into the environment and to minimize other environmental hazards.

The 1978 act provides that

1. In cooperation with the United States, a program of assessment and remedial action for mill tailing sites will include, when appropriate, the reprocessing of tailings to extract residual uranium and other minerals in order to stabilize and control such tailings in a safe and environmentally sound manner and to minimize and eliminate radiation health hazards to the public.
2. A program be established to regulate mill tailings during uranium or thorium ore processing at active mill operations and after termination of such operations in order to stabilize and control such tailings in a safe and environmentally sound manner and to minimize or eliminate radiation health hazards to the public.

The act identified 20 processing sites containing residual radioactive materials at which all or substantially all of the uranium was produced for sale to federal agencies prior to January 1, 1971, and other sites that the Nuclear Regulatory Commission has designated to be contaminated with residual radioactive materials. The designated processing sites were:

Salt Lake, Utah
Green River, Utah
Mexican Hat, Utah
Durango, Colorado
Grand Junction, Colorado
Rifle, Colorado (two sites)
Gunnison, Colorado
Naturita, Colorado
Maybell, Colorado
Slick Rock, Colorado (two sites)
Shiprock, New Mexico
Ambrosia Lake, New Mexico
Riverton, Wyoming
Lakeview, Oregon
Falls City, Texas
Tuba City, Arizona
Monument Valley, Arizona
Lowman, Idaho
Canonsburg, Pennsylvania

To implement the act, the secretary of energy was designated to develop cooperative agreements with the states to perform remedial actions at each processing site. Each cooperative agreement required the state to meet the conditions of the secretary of energy. In a designated processing area, the secretary of energy was required to pay 90 percent of the cost of the remedial action, including the cost of the site. If the provisions of the act were violated, the secretary of energy could impose a penalty of not more than $1,000 per day per violation.

The act also required reports to Congress to include:

1. Data on the actual and estimated costs of the program authorized
2. A description of the extent of participation by the states and Indian tribes in this program
3. An evaluation of the effectiveness of remedial actions and a description of any problems associated with the performance of such actions
4. Such other information as might be appropriate.

Human Health and Safety

Radiologic procedures are increasingly used to diagnose and treat diseases. Although these procedures have positive results, there is also a danger of radiation contamination. Congress has recognized that controls are necessary. To illustrate these efforts, two quite different acts have been enacted. The Consumer-Patient Radiation Health and Safety Act of 1981 provides guidelines for the use of radiologic procedures. The Nuclear Protection and Safety Act of 1987 establishes procedures for health and safety at facilities owned by the Department of Energy that have had self-regulation in the past.

Consumer-Patient Radiation Health and Safety Act of 1981

It has long been recognized that many medical procedures that utilize radiologic procedures can be hazardous to health. In the passage of this act Congress stated that

1. It is in the interest of public health and safety to minimize unnecessary exposure to potentially hazardous radiation due to medical and dental radiologic procedures.

2. It is in the interest of public health and safety to have a continuing supply of adequately educated persons and appropriate accreditation and certification programs administered by state governments.
3. The protection of the public health and safety from unnecessary exposure to potentially hazardous radiation due to medical and dental radiologic procedures and the assurance of efficacious procedures are the responsibility of state and federal governments.
4. Persons who administer radiologic procedures, including those done at federal facilities, should be required to demonstrate competence by reason of education, training, and experience.
5. The administration of radiologic procedures and their effect on individuals have a substantial and direct effect upon United States interstate commerce.

The purpose of the act is to

1. Provide for the establishment of minimum standards by the federal government for the accreditation of education programs for persons who administer radiologic procedures and for the certification of such persons
2. Ensure that medical and dental radiologic procedures are consistent with rigorous safety precautions and standards

The act indicated that standards were to be established for

1. Medical radiologic technologists
2. Dental auxiliaries
3. Radiation therapy technologists
4. Nuclear medicine technologists
5. Other auxiliaries who administer radiologic procedures

The act also indicated that the secretary of health and human services, along with other agencies, would prepare federal radiation guidelines with respect to radiologic procedures. Such guidelines shall

1. Determine the level of unnecessary radiation exposure from radiologic procedures and specify techniques, procedures, and methods to minimize such unnecessary exposure
2. Provide for the elimination of the need for retakes of diagnostic radiologic procedures

3. Provide for the elimination of unproductive screening programs
4. Provide for the optimum diagnostic information with minimum radiologic exposure
5. Include the therapeutic application of radiation to individuals in the treatment of disease, including nuclear medicine applications.

Nuclear Protection and Safety Act of 1987

Since the beginning of the nuclear age, the Department of Energy (DOE) and its predecessor agencies have not been subject to outside, independent oversight of their technical operations. Originally this policy may have been justified on the basis of national security. In recent years it has become evident that the original policy was not satisfactory. While it is the responsibility of DOE to produce nuclear materials for weapons as well as for nonmilitary purposes, DOE also has the responsibility for ensuring health and safety at the facilities without being subject to appropriate and independent outside review. Thus, DOE facilities operate without the full panoply of federal protection and independent oversight found in the commercial nuclear industry.

Today the DOE nuclear complex includes 127 facilities in 23 states employing over 142,000 people. The facilities range in size from a single building with one employee to the giant Savannah River Plant in South Carolina and the Hanford Reservation in Washington, each with over 10,000 employees. After World War II all commercial plants had to be licensed by the Atomic Energy Commission. However, in the tense cold war atmosphere and the concern for technological security, the DOE member facilities were exempted from licensing requirements.

The record shows that many problems have developed during DOE's long history of self-regulation. In 1987 the Committee on Governmental Affairs hearings indicated:

> The General Accounting Office work on safety matters at DOE facilities over the last award years . . . has identified important safety issues, which considering their scope, raise serious questions about both the safety of individual facilities and DOE operations as a whole.

Because the DOE nuclear complex includes many aged facilities, the potential for additional safety problems is increased. For example, the last construction of a reactor was in 1963.

The purpose of the Nuclear Protection and Safety Act of 1987 was:

1. To establish an independent Nuclear Safety Board to reduce oversights and to promote safety of the DOE's nuclear facilities
2. To review and evaluate the implementation of current health and safety standards and DOE orders at DOE nuclear facilities
3. To investigate accidents that affect health and safety
4. To recommend changes in operating procedures to improve health or safety standards
5. To make recommendations to ensure that standards of design and construction be commensurate with standards in private-sector facilities.

To implement health and safety standards the Occupational Safety and Health Administration (OSHA) and the National Institute for Occupational Safety and Health (NIOSA) will assess the health and safety of DOE facilities.

International Control

In 1945, with the development and initial explosion of the atomic bomb in Japan, it was recognized by the countries capable of producing nuclear weapons that this new source of destruction had to come under international control. It was also recognized that atomic energy should be used to serve mankind rather than destroy it. As a consequence a long and difficult struggle began in international diplomacy, one that has no end in sight. The struggle between the great powers, primarily the Soviet Union and the United States, continues to limit the use of atomic weapons. At the international level the nations agree that the United Nations is the instrument to provide world leadership in these endeavors. Whether national and international goals can be reconciled remains a major unanswered question.

Agreed Declaration of the President of the United States, the Prime Minister of the United Kingdom, and the Prime Minister of Canada

This declaration was prepared on November 15, 1945. It recognized that

1. The application of recent scientific discoveries to the methods and practice of war has placed at the disposal of mankind means of destruction hitherto unknown.
2. The responsibility for devising means to ensure that the new discoveries shall be used for the benefit of mankind, instead of as a means of destruction, rests not on our nations alone, but upon the whole civilized world.
3. Representing as we do, the three countries which possess the knowledge essential to the use of atomic energy, we declare at the outset our willingness, as a first contribution, to proceed with the exchange of fundamental scientific information and the interchange of scientists and scientific literature for peaceful ends with any nation that will fully reciprocate.
4. In order to attain the most effective means of entirely eliminating the use of atomic energy for destructive purposes and promoting its widest use for industrial and humanitarian purposes, we are of the opinion that at the earliest practicable date a Commission should be set up under the United Nations Organization to prepare recommendations for submission to the Organization.

The commission was to make specific proposals as to

1. Extending between all nations the exchange of basic scientific information for peaceful ends
2. Control of atomic energy to the extent necessary to ensure its use only for peaceful purposes
3. The elimination from national armaments of atomic weapons and all other major weapons adaptable to mass destruction
4. Effective safeguards by way of inspection and other measures to protect complying states against the hazards of revolutions and invasions

The declaration was signed by Harry S. Truman, C. R. Attlee, and W. L. Mackenzie King.

Moscow Conference of the Three Foreign Ministers of the United States, United Kingdom, and the Soviet Union

On December 27, 1945, the foreign ministers of the United States, the United Kingdom, and the Soviet Union agreed to recommend that at the first session of the General Assembly of the United Nations in January 1946 a Commission for the Control of Atomic Energy be established. The commission was to be composed of one representative from each of the nations represented on the Security Council, plus Canada. The purpose of the commission was the same as proposed above by the United States, United Kingdom, and Canada. The report was signed by James F. Byrnes, Ernest Bevin, and V. Molotov.

Early Attempts at International Regulation

The United States sought international agreement for placing nuclear materials and activities relevant to nuclear weaponry under the aegis of the United Nations. This concept was based on the viewpoint that the dangers to international security were so great due to the atomic bomb that individual nations should be denied the sovereign rights to it. This was initially known as the Acheson-Lilienthal Plan but later became known as the Baruch Plan after Bernard Baruch presented it to the United Nations in 1946.

The United States proposed the creation of an International Atomic Development Authority, to which would be given responsibility for all aspects of the development of atomic energy, including:

1. Managerial control or ownership of all atomic energy activities potentially dangerous to world security
2. Power to control, inspect, and license all other atomic activities
3. The duty of fostering the beneficial areas of atomic energy
4. Research and development responsibilities of an affirmative character intended to put the authority in the forefront of atomic knowledge and thus enable it to comprehend, and therefore to detect, misuse of atomic energy.

This plan was based on the premise that with the development of nuclear energy for peaceful activities there would also be the acceptance of arms control measures, including inspection or safeguarding by foreign nationals. Immediately, disagreement developed between the United States and the Soviet Union. The United States insisted the authority be founded before it gave up its nuclear weapons; the Soviets insisted the opposite should apply. Because neither country trusted the other, the plan was not implemented.

The U.S. policy on development and use of atomic energy was legislated in the Atomic Energy Act of 1946. This plan, in operation to 1953, denied all other countries access to knowledge, technology, and materials that had a bearing on nuclear weapons. In addition, the scientific community of the United States was prevented from freely conducting nuclear research or communicating its results. Nevertheless, this policy of scientific secrecy did not prevent the first wave of weapons proliferation—the Soviet Union was first to develop a bomb, followed by the United Kingdom and France.

This early experience demonstrated a central point about a nonproliferation policy. Measures to control the spread of nuclear material and technology have essentially no chance of success if the political and strategic environment is such that a nation desires nuclear weaponry for its security.

Soon after the Soviets exploded their first atomic bomb in 1949, President Truman announced the United States would develop the still more potent thermonuclear, or hydrogen, bomb. The first U.S. H-bomb test in 1951 was matched by the Soviets. This demonstrated that the Soviet nuclear capability was dynamic. A general policy of denying the Soviets technical information did not prevent the development of the weapons. The challenge by the middle 1950s was not how to stop the spread of nuclear weapons, but how each nation could surpass the other in a nuclear arms race.

The United States became concerned that the Soviet Union would begin to supply nuclear technology to other countries, particularly Third World nations. In order to counteract this, President Eisenhower announced the Atoms for Peace program in July 1953 at a meeting of the United Nations General Assembly. Atoms for Peace was one of the first endeavors in the cold war to neutralize the efforts of the Soviets. The pressure to develop nuclear technology for peaceful endeavors was intensified by predictions that nuclear power would precipitate a new industrial revolution.

Treaty on the Non-Proliferation of Nuclear Weapons (NPT)

On July 1, 1968, the governments of the Soviet Union, the United States, and the United Kingdom (known as parties to the treaty) signed the Treaty on the Non-Proliferation of Nuclear Weapons. The preamble to this nonproliferation treaty stated:

1. Considering the devastation that would be visited upon all mankind by a nuclear war and the consequent need to make every effort to avert the danger of such a war and to take measures to safeguard the security of people,
2. Believing that the proliferation of nuclear weapons would seriously enhance the danger of nuclear war,
3. In conformity with resolutions of the United Nations General Assembly calling for the conclusion of an agreement on the prevention of wider dissemination of nuclear weapons,
4. Undertake to cooperate in facilitation the application of International Atomic Energy Agency safeguards on peaceful nuclear activities,
5. Expressing their support for research, development and other efforts to further the application, within the framework of the International Atomic Energy Agency safeguards system, of the principle of safeguarding effectively the flow of source and special fissionable materials by use of instruments and other techniques at certain strategic points,
6. Affirming the principle that the benefits of peaceful applications of nuclear technology, including any technological by-products which may be derived by nuclear-weapon States from the development of nuclear explosive devices, should be available for peaceful purposes to all Parties to the Treaty, whether nuclear-weapon or non-nuclear-weapon States,
7. Convinced that, in furtherance of this principle, all Parties to the Treaty are entitled to participate in the fullest possible exchange of scientific information for, and to contribute alone or in cooperation with other States to the further development of the applications of atomic energy for peaceful purposes,
8. Declararing their intention to achieve at the earliest possible date the cessation of the nuclear arms race and to undertake effective measures in the direction of nuclear disarmament,
9. Urging the cooperation of all States in the attainment of this objective,
10. Recalling the determination expressed by the Parties to the 1963 Treaty barring nuclear weapon tests in the atmosphere, in outer space and under water in the Preamble to seek to achieve the discontinuance of all test explosions of nuclear weapons for all time and to continue negotiations to this end,

11. Desiring to further the easing of international tension and the strengthening of trust between States in order to facilitate the cessation of the manufacture of nuclear weapons, the liquidation of all their existing stockpiles, and the elimination from national arsenals of nuclear weapons and the means of their delivery pursuant to a treaty on general and complete disarmament under strict and effective international control,
12. Recalling that, in accordance with the charter of the United Nations, States must refrain in their international relations from the threat or use of force against the territorial integrity or political independence of any State, or in any other manner inconsistent with the Purposes of the United Nations, and that the establishment and maintenance of international peace and security are to be promoted with the least diversion for armaments of the world's human and economic resources.

Have agreed as follows:

1. Each nuclear-weapon State Party to the Treaty undertakes not to transfer to any recipient nuclear weapons or explosives to control weapons and explosive devices directly or indirectly, or the manufacturing of such weapons.
2. Each non-nuclear-weapon State Party to the Treaty will not receive nuclear weapons or explosives, or manufacture weapons.
3. Each State Party to the Treaty undertakes not to provide (a) source or special fissionable material, or (b) equipment or material used to produce fissionable material.
4. Nothing in the Treaty will affect the right of the Parties to the Treaty to develop research, production and use of nuclear energy for peaceful purposes.
5. Each of the Parties of the Treaty pursue negotiations for the cessation of the nuclear arms race at an early date and ultimately to nuclear disarmament.
6. Regional treaties may be negotiated to assure the disappearance of nuclear weapons in their respective territories.
7. The treaty is open to all States for signature.
8. A State may withdraw from the Treaty if the supreme interests of the country are jeopardized.
9. Twenty-five years after the entry into force of the Treaty, a conference shall be convened to decide whether the Treaty shall continue in force indefinitely, or shall be extended for an additional fixed period or periods.

The treaty has now been ratified by at least 128 countries and signed but not ratified by 3 more. Thirty-three nations have not ratified the treaty, including the nuclear weapons nations of France, China, and India. Other major nations not ratifying the treaty are Brazil, Chile, Cuba, Israel, Pakistan, Saudi Arabia, and Spain.

Nuclear Non-Proliferation Act of 1978

Congress by this act established the policy that nuclear explosives posed a grave threat to the security of the United States and continued international programs toward world peace and development. Because recent events emphasize the urgency of this threat, there is an imperative need to increase the effectiveness of international safeguards and controls on peaceful nuclear activities to prevent proliferation.

The stated policy of the United States is to:

1. Actively pursue through international initiatives mechanisms for fuel supply assurances and the establishment of more effective international controls over the transfer and use of nuclear materials, equipment and nuclear technology for peaceful purposes in order to prevent proliferation, including the establishment of common international sanctions.
2. Take such actions as are required to confirm the reliability of the United States in meeting its commitments to supply nuclear reactors and fuel to nations which adhere to effective non-proliferation policies.
3. Strongly encourage nations which have not ratified the Treaty on the Non-Proliferation of Nuclear Weapons, to ratify it.
4. Cooperate with foreign nations in identifying and adopting suitable technologies for energy production and, in particular, to identify alternative options to nuclear power in aiding such nations to meet their energy needs, consistent with the economic and material resources of those nations.

To implement the policies of the act agrees to:

1. Establish a more effective framework for international cooperation to meet the energy needs of all nations and to ensure that the worldwide development of peaceful nuclear activities and the export by any nation of nuclear materials and equipment and nuclear technology intended for use in peaceful nuclear activities do not contribute to proliferation of weapons.
2. Authorize the United States to ensure that it will act reliably in meeting its commitment to supply nuclear reactors and fuel to nations which adhere to effective nonproliferation policies.

3. Provide incentives to other nations of the world to join international cooperation efforts.
4. Ensure effective control by the United States over its export of nuclear materials, equipment and technology.

The act has six titles. Title 1 stipulates the policies for providing an adequate nuclear fuel supply. Specifically, it calls for an expansion in the uranium enrichment capacity and a reevaluation of the nuclear fuel cycle. Title 2 stresses initiatives to strengthen the international safeguards through the Treaty on the Non-Proliferation of Nuclear Weapons and the International Atomic Energy Policy. A training program is to be established by the Department of Energy with the Nuclear Regulatory Commission to train persons from nations that have developed or expect to develop nuclear materials and equipment.

Titles 3 and 4 consider the export organization with an emphasis on licensing procedures, standards, protection of trade secrets, and transfer procedures. Title 5 considers the United States' assistance to developing countries, including energy needs for development, reduction of such countries' dependence on petroleum fuels, with emphasis given to utilizing solar and other renewable energy resources, and expanding alternative means of energy.

The Treaty of Tlatelolco

The establishment of nuclear-weapon-free zones has been advocated for many areas of the world. International prohibitions have been approved for the emplacement of nuclear weapons on the ocean floor and the stationing or orbiting of those weapons in space or on celestial bodies. Antarctica as a demilitarized area is ipso facto a nuclear-weapon-free zone. The Treaty of Tlatelolco for the prohibition of nuclear weapons in Latin America is the only treaty so far, however, establishing a nuclear-weapon-free zone among a group of nations.

The Treaty of Tlatelolco was completed for ratification on February 14, 1967, in Mexico City. It was the first agreement to establish a system of international controls and a permanent supervisory organ, the Agency for the Prohibition of Nuclear Weapons in Latin America (OPAWAL). The treaty consists of 31 articles, one transitional article, and two additional protocols that establish a system of mutual rights and obligations. The treaty was initiated in 1962 by Bolivia, Brazil, Chile, and Ecuador. In 1967 the treaty was accepted by the United Nations General

Assembly, which recommended that all Latin American countries adhere to the treaty's provisions. Parties to the treaty are required to enter into agreement with the International Atomic Energy Agency (IAEA). The treaty also creates a system of inspections to deal with suspected cases of violation.

Article 1 establishes the framework that atomic energy will be used exclusively for peaceful purposes and that there will be no:

1. Testing, use, manufacture, production, or acquisition by any means whatsoever of any nuclear weapons, by the parties themselves, directly or indirectly, on behalf of anyone else or in any other way, and
2. The receipt, storage, installation, deployment and any form of possession of any nuclear weapons, directly or indirectly, by the parties themselves, by anyone on their behalf, or in any other way.

Article 5 defines a nuclear weapon as

> Any device which is capable of releasing nuclear energy in an uncontrolled manner and which has a group of characteristics that are appropriate for use for warlike purposes.

Article 18 provides the rights of the contracting parties to carry out nuclear explosions for peaceful purposes, including explosions which involve devices similar to those used in nuclear weapons, provided that those explosions are carried out in accordance with the treaty. This provision has given rise to some possible complications. In signing the treaty, Argentina, Brazil, and Nicaragua have reserved the right to carry out peaceful nuclear explosions (PNEs). Nicaragua has stated that it reserves

> Its sovereign right to use nuclear energy, as it deems fit, for such peaceful purposes as the removal of large amounts of earth for the construction of interoceanic or any other type of canal, irrigation works, electric power station, etc. and to allow the transit of atomic materials through its territory.

In contrast, some countries interpret the treaty as prohibiting the use of nuclear explosive devices for peaceful purposes unless and until nuclear devices are developed that can be distinguished from nuclear weapons. The United States, the United Kingdom, and the Soviet Union have declared that they are in complete opposition to the use of any nuclear explosive devices in Latin America.

Minor items of controversy continue over implementation of the treaty. Differences exist concerning the geographical extent of the weapons-free zone. According to the treaty, its general application embraces the territory, territorial sea, air space, and any other space over which each of the zonal states exercise sovereignty in accordance with its own legislation. Such legislation, however, varies from state to state.

Another source of controversy with regard to the treaty is unsettled territorial disputes, such as between Venezuela and Guyana and Guatemala and Belize. An additional difference exists on the question of transit. The treaty does not specifically prohibit either transit or transport of nuclear weapons. The issue is viewed by the nations in different ways. For example, Mexico and Panama have indicated they will not permit transit of nuclear weapons through their territories.

As of May 1, 1985, 23 Latin American countries had signed the Treaty of Tlatelolco. The treaty also includes the territories of the Netherlands, the United Kingdom, and the United States. Brazil and Chile have ratified the treaty but have not waived the requirements. Argentina has signed the treaty and has stated it shares its aims and objectives. Belize and Guyana had not been invited by the General Conference to ratify the treaty because of territorial disputes. Five states, Cuba, Dominica, Saint Lucia, Saint Christopher and Nevis, and Saint Vincent and the Grenadines, had not signed the treaty. The last four are newly created states and are expected to sign the treaty.

Twenty-one Latin American states are members of both the Treaty of Tlatelolco and the Treaty on the Non-Proliferation of Nuclear Weapons (NPT). Most states see no conflict between these two treaties. A few states consider the NPT discriminatory. These include Argentina, Brazil, Chile, Cuba, and Guyana. The Soviet Union, the United Kingdom, and the United States believe that the NPT is a cornerstone of international efforts to erect and to sustain effective barriers to further spread of nuclear weapons. These states have made it clear that acceptance of NPT is a necessary first step for the establishment of a nuclear-weapon-free zone.

4

Directory of Organizations

A LARGE NUMBER of governmental and private organizations have been created to consider the nuclear industry. In the United States the Nuclear Regulatory Commission is the major governmental organization. There are also a large number of governmental advisory committees.

The private organizations vary from those having a wide variety of functions to those dedicated to a single objective. Most of the organizations are permanent, but because organizational groups have increased rapidly in the past decade, some may discontinue functioning when their objectives are completed. There are also international organizations. Many of these are located in Europe and have a specific purpose.

The following directory is arranged in three sections: Private Organizations in the United States, U.S. Government Organizations, and International Organizations.

Private Organizations in the United States

American Nuclear Energy Council (ANEC)
410 First Street, SE
Washington, DC 20003

Established: 1975. Members: 125. Staff: 15.

Purpose: Supports the development of nuclear power as an energy source and coordinate and project the interests of the American nuclear industry to Congress and the executive branch.

Activities: Relates congressional and executive branch actions affecting nuclear energy issues to member companies.

Publications: None

American Nuclear Society (ANS)
555 N. Kensington Avenue
La Grange Park, IL 60525

Established: 1954. Members: 14,000. Staff: 86. Local Groups: 51 (9 overseas). Student Branches: 53. Semiannual meetings.

Purpose: To advance science and engineering in the nuclear industry, disseminate information, promote research, and establish scholarships.

Activities: Conducts meetings devoted to scientific and technical papers, work with governmental agencies, educational institutions, and other organizations dealing with nuclear issues. Maintains a library of 3,000 volumes on nuclear emergency and business management.

Committees: Engineering and Technology Accreditation, Registration and Professional Development, Honors and Awards, International, Nuclear Engineering Education for the Disadvantaged, Professional Divisions, Public Information.

Publications:
ANS News, monthly
Nuclear News, monthly
Nuclear Science and Engineering, monthly
Nuclear Standard News, monthly
Nuclear Technology, monthly
Fusion Technology, bimonthly
Transactions, semiannual
ANS/DOE monograph series

Americans for Nuclear Energy (AFNE)
2525 Wilson Boulevard
Arlington, VA 22201

Established: 1978. Members: 18,000. Staff: 1.

Purpose: Citizens in favor of nuclear energy.

Activities: Lobbies Congress and state legislatures. Rates Congress on nuclear issues. Enters lawsuits and favors election issues that support nuclear power. Provides speakers. Sponsors citizen conferences.

Publications:
Nuclear Advocate, monthly
Alpha Series, pamphlets

Asian and Pacific Americans for Nuclear Awareness (APANA)
6181 Sylvan Drive
Santa Susana, CA 93063

Established: 1980. Members: 100. Annual Hiroshima/Nagasaki commemoration. Periodic conferences.

Purpose. To educate the public about nuclear war, disarmament, and especially the history of those who survived the atomic bombings of Hiroshima and Nagasaki.

Activities: Conducts research on a nuclear-free Pacific. Compiles statistics, bestows awards, operates speakers' bureau. Developing a curriculum for classroom use to establish world peace.

Publications:
Newspaper, periodic
Brochure

Citizen's Energy Council (CEC)
77 Homewood Avenue
Allendale, NJ 07401

Established: 1966. Members: 2,200.

Purpose: Inform community leaders concerned with the hazards of nuclear energy.

Activities: Monitors and reports on activities of the Nuclear Regulatory Commission and other events related to nuclear energy. Conducts training sessions on community organization, organizes workshops. Has a 420-volume library on the history of the movement to oppose nuclear power.

Publications:
Energy News Digest (newspaper), monthly
Last Chance Bulletin, bimonthly

Concerned Educators Allied for a Safe Environment (CEASE)
17 Gerry Street
Cambridge, MA 02138

Established: 1979. Members: 577. Annual meeting.

Purpose: Seeks to inform the public on issues involving nuclear warfare and peace alternatives. Works to aid in the redirection of funds allocated for the military to develop human resources.

Activities: Sponsors workshops in cooperation with National Association for Education of Young Children. Informs public on issues of nuclear warfare.

Publications:
CEASE News, quarterly

Environmental Coalition on Nuclear Power (ECNP)
433 Orlando Avenue
State College, PA 16803

Established: 1970. Meetings: quarterly.

Purpose: Seeks establishment of a safe, nonnuclear energy policy.

Activities: Conducts many educational, legal, and political activities at local, state, and national levels. Has a 2,000-volume library on the Leo Goodman collection.

Publications:
Newsletter (periodic)

Fusion Energy Foundation (FEF)
Address unknown

Established: 1974. Members: 20,000. Staff: 40.

Purpose: To provide a forum of independent, high-level scientific discussion of fusion from the standpoint of comprehensive policy making.

Activities: Administers research programs and publishes articles concerned with historical aspects of the frontiers of science. Holds national and regional forums, seminars, and conferences. Provides speakers on nuclear and high-technology development, economic development, and development of directed energy beam technologies for defensive weapons.

Publications:
Fusion, bimonthly (in English, French, German, Italian, Spanish)
International Journal of Fusion Energy, quarterly
Special reports

Health and Energy Institute (HEI)
236 Massachusetts Avenue, NE, No. 506
Washington, DC 20002

Established: 1978. Members: 7,000. Semiannual Radiation Victims Roundtable. Periodic conferences.

Purpose: To inform about the impact of nuclear energy on health and the environment with emphasis on radiation danger.

Activities: Conducts research and educational programs geared to the general public on the effects of food irradiation. Provides speakers for high schools. Sponsors National Network to Prevent Birth Defects. Maintains library.

Publications:
Radiation Victims Organization Directory, periodic
Brochures

Institute of Nuclear Materials Management (INMM)
60 Revere Drive, Suite 500
Northbrook, IL 60062

Established: 1958. Members: 800. Annual meeting.

Purpose: To promote the principles of science, business, and industry to nuclear materials management.

Activities: Sponsors educational courses in nuclear materials control and safeguards. Sponsors American National Standards Institute committee and Standards for the Control of Nuclear Materials. Presents Distinguished Service Award and Meritorious Service Award. Maintains a speakers' bureau.

Publications:
Journal of Nuclear Materials Management, quarterly

Institute of Nuclear Power Operation (INPO)
1100 Circle 75 Parkway, Suite 1500
Atlanta, GA 30339

Established: 1979. Members: 101. Staff: 440.

Purpose: Aids in the development of plant operations and develops criteria.

Activities: Electric utilities operating nuclear power plants. Operates the National Academy for Nuclear Training. Conducts workshops and conferences for various levels of operations staff and management. Provides scholarships and fellowships to schools of engineering. Maintains a 10,000-volume library.

Publications:
Nuclear Professional, semiannual
Review, semiannual

Institute for Space and Security Studies (ISSS)
7833 C Street
Chesapeake Beach, MD 20732

Established: 1983. Members: 26,000. Periodic meetings.

Purpose: Seeks to teach the public and Congress about nuclear weapons systems. Goal is to prevent nuclear war.

Activities: Sponsors workshops, news conferences, congressional briefings, provides educational video, sponsors graduate assistantships, and supports research fellows and interns. Maintains a library.

Publications:
Space and Security News, quarterly
Annual Report
Brochures

National Organization of Test, Research and Training Reactors (NOTRTR)
Radiation Center
Oregon State University
Corvallis, OR 97331

Established: 1972. Members: 100. Annual conference.

Purpose: Forum on the utilization and operation of nonpower nuclear reactor facilities.

Activities: Conducts informative exchange programs for members. Studies government rules and regulations concerning nonpower facilities. Maintains a speakers' bureau, compiles statistics.

Publications:
Directory, periodic

Nuclear Energy Women (NEW)
1776 I Street, NW, Suite 4000
Washington, DC 20006

Established: 1975. Members: 500. Regional groups: 15. State groups: 7. Annual meeting.

Purpose: Individuals in energy industries and citizens advocacy groups providing educational services.

Activities: Disseminates nuclear energy information to women's groups and others. Seeks to provide an understanding of energy choices. Goal is to establish a national network of women with energy expertise who will foster energy education. Develops and conducts national and regional energy workshops and seminars. Arranges energy tours. Bestows Outstanding Woman in Energy Award annually.

Publications: None

Nuclear Information and Records Management Association (NIRMA)
210 Fifth Avenue
New York, NY 10010

Established: 1978. Members: 415. Staff available. Annual meeting.

Purpose: To advance the theory and practices of management of corporate records primarily pertaining to nuclear facilities through research, preparedness of papers, and reports on practices in the collection, storage, maintenance, and retrieval of nuclear records.

Activities: Committees conduct research on a continuing basis. Maintains a library of symposium proceedings, position papers, and guides.

Publications:
Newsletter, quarterly

Professional Reactor Operator Society (PROS) (Nuclear)
P.O. Box 181
Mishicot, WI 54228

Established: 1981. Members: 700. Regional groups. Annual conference.

Purpose: To develop a communication network between nuclear reactor operators and government agencies, Congress, and industry in order to promote safety and efficiency in nuclear facilities.

Activities: Survey of views and concerns of members and other involved parties. Provides guidelines for training of reactor personnel.

Publications:
The Communicator (newsletter), quarterly

Society for the Advancement of Fission Energy (SAFE)
336 Coleman Drive
Monroeville, PA 15146

Established: 1976. Members: 1,200. Regional groups.

Purpose: To encourage the use of nuclear energy to supply electricity.

Activities: Conducts seminars, publishes a "state's list" for citizens interested in forming grass roots energy organizations. Bestows Black Bulb Award to politicians who delay nuclear power advancements. Presently inactive.

Publications: None

Universities Research Association (URA) (Nuclear)
1111 19th Street, NW, Suite 400
Washington, DC 20036

Established: 1965. Members: 56. Council of Presidents in which each member university is represented by its chief officer. Regional groups. Annual meeting.

Purpose: To constitute an entity in and by means of which universities and other research organizations may cooperate with one another, with the government of the United States, and with other organizations.

Activities: Under contract to the Department of Energy to operate the Fermi National Accelerator Laboratory in Illinois.

Publications:
Annual report

Utility Nuclear Waste Management Group (UNWMG)
1111 19th Street, NW
Washington, DC 20036

Established: 1980. Members: 43. Staff: 5.

Purpose: Electric utilities united to promote solutions to the problems of radioactive waste management.

Activities: Seeks to monitor high-level and low-level nuclear waste management programs. Maintains task forces and working groups for specific problems.

Publications: None

U.S. Government Organizations

Government Commission

Nuclear Regulatory Commission (NRC)
1717 H Street, NW
Washington, DC 20555

Established: Commission established as an independent regulatory agency under the provisions of the Energy Reorganization Act of 1974 and Executive Order of January 15, 1975.

Programs: The major programs of the NRC are the Office of Nuclear Reactor Regulation, the Office of Nuclear Material Safety and Safeguards, and the Office of Nuclear Regulatory Research. These programs were created by the Energy Reorganization Act of 1974.

Purpose: The purpose of the Nuclear Regulatory Commission is to ensure that the civilian uses of nuclear materials and facilities are conducted in a manner consistent with the public health and safety, environmental quality, national security, and the antitrust laws. The major share of the commission effort is focused on regulating the use of atomic energy to generate electricity.

Activities: The NRC fulfills its obligations through a system of licensing and regulations that includes:

1. Licensing the construction and operation of nuclear reactors and other nuclear facilities and the possession, use, processing, handling, and disposal of nuclear materials
2. Regulation of licensed activities, including assurance that measures are taken for the physical protection of facilities and materials
3. Development and implementation of rules and regulations governing licensed nuclear activities
4. Inspection of licensed facilities and activities
5. Investigation of nuclear incidents and allegations concerning any matters relating to the NRC
6. Enforcement of the NRC licenses and regulations by the issuance of orders, civil penalties, and other types of actions, including public hearings
7. Conducting public hearings on nuclear and radiological safety, environmental concerns, common defense and security, and antitrust matters
8. Development of effective working relationships with the states regarding the regulation of nuclear materials. The relationship includes the assurance that adequate regulatory programs are maintained by those states that exercise by agreement with the commission regulatory control over certain nuclear materials within their respective borders. In addition, a systematic review of operational data, including reports of accidents and other events, from nuclear power plants is performed in order to detect trends that will better enable NRC to forecast and solve safety problems.

Employment: NRC's employment policies are exempt from civil service requirements and are conducted under NRC's independent merit system.

Publications:

Annual report provides a summary of major activities for the year.

NRC produces a variety of scientific, technical, and administrative bulletins dealing with licensing and regulating civilian nuclear power. NRC maintains its principal public document room at 1717 H Street NW, Washington, DC. The main public document room contains about 1.4 million documents in hard copy and microfiche. These documents pertain to the licensing of source material, production and utilization facilities, special nuclear materials, transportation of radioactive materials, research and technical assistance reports, reports on technical matters, rules and regulations, commission correspondence, and other items. The library is open to the public.

In addition, NRC has about 100 local public document rooms around the country. They are typically located in libraries in cities and towns near nuclear power plants and contain information specific to the nearby facility.

Government Advisory Committees

Ad Hoc Advisory Committee for the Review of Enforcement Policy
(Nuclear Safety)
Physics Division
Directorate for Mathematical and Physical Sciences
National Science Foundation
1800 G Street, NW
Washington, DC 20550

Established: Committee was established September 5, 1984, as an ad hoc public advisory committee of the Nuclear Regulatory Commission. Membership: 5. Meetings: 4 to 6 yearly.

Purpose: To review NRC enforcement policies.

Activities: The activities of the committee include the promotion and protection of the radiological health and safety of the public, including employees' health and safety, the common defense and security, and the environment, and to make recommendations for changes in the enforcement policy.

Publications: None

Advisory Committee on Reactor Safeguards (Nuclear Safety)
Nuclear Regulatory Commission
Washington, DC 20555

Established: Committee was established in 1957 by Sections 29(b) and 182 of the Atomic Energy Act of 1954, as amended. Membership: Consists of 15 members appointed by the NRC for terms of 4 years. Meetings: About 85 meetings are held each year.

Purpose: This is a public advisory committee that reports directly to the Nuclear Regulatory Commission on reactor safeguards.

Activities: Committee reviews safety studies and facility license applications and prepares reports; advises the commission with regard to the hazards of proposed or existing reactor facilities and the adequacy of proposed reactor safety standards. Committee, as required by P.L. 95-209, prepares an annual report to Congress on the NRC Safety Research Program.

Publications:
Numerous reports available from the NRC on request.

Advisory Panel for the Decontamination of Three Mile Island, Unit 2
Three Mile Island Cleanup Project Directorate
Nuclear Regulatory Commission
Washington, DC 20555

Established: Panel established November 18, 1980, and terminated November 1981 and reestablished November 28, 1986.

Purpose: Consults and advises the Nuclear Regulatory Commission on major activities required.

Activities: Studies the length of time necessary to bring the reactor under control and if there was a major release in radioactivity.

Publications: None

Atomic Safety and Licensing Board Panel (Nuclear Energy)
Nuclear Regulatory Commission
7910 Woodmont Avenue
Washington, DC 20545

Established: Panel was established under authority of the Atomic Energy Act, as amended by P.L. 87-615 and the Energy Reorganization Act of 1974. Membership: Panel includes 21 full-time members plus 44 appointed members, including persons qualified in administrative and antitrust law, nuclear physics, chemistry, engineering, economics, and environmental disciplines. Meetings: Full panel meets once a year as a judicial body. Other meetings as appropriate.

Purpose: To provide safety standards and to license atomic facilities.

Activities: Atomic safety and licensing boards are selected from the panel. Each three-person board conducts reviews and hearings on the application for the construction and operation of civilian nuclear power plants. The board is concerned with safety, environmental, and antitrust aspects.

Publications: None

Central Interstate Low-Level Radioactive Waste Commission
(Nuclear Waste)
Office of Air Quality and Nuclear Energy
Department of Environmental Quality
P.O. Box 14690
Baton Rouge, LA 70898

Established: Commission was established under authority of the Central Interstate Low-Level Radioactive Waste Compact, Section 222 of Title II of P.L. 99-240, the Omnibus Low-Level Radioactive Waste Interstate Compact Central Act of January 15, 1986. Membership: Commission comprises one member from each state party to the compact. Meetings: Commission meets at least once a year.

Purpose: Commission was established to initiate proceedings or act as intervenor before any court of law or any federal, state, or local agency, board, or commission that has jurisdiction of disposal of low-level radioactive waste.

Activities: Compact provides for the efficient and economic management of low-level waste and for the health, safety, and welfare of the residents of the states associated with the compact. States in the compact include Arkansas, Iowa, Kansas, Louisiana, Minnesota, Missouri, Nebraska, North Dakota, and Oklahoma.

Publications: None

Central Midwest Interstate Low-Level Radioactive Waste Commission
(Nuclear Waste)
Illinois Department of Nuclear Safety
1035 Outer Park Drive
Springfield, IL 62704

Established: Commission was established under authority of the Central Midwest Interstate Low-Level Radioactive Waste Compact, Section 224 of Title II of P.L. 99-240, the Omnibus Low-Level Radioactive Waste Interstate Compact Consent Act of January 15, 1986. Membership: Two members from Illinois and one from Kentucky. Meetings: At least one meeting every year.

Purpose: Commission was established to enter into agreement with any person, state, or group of states for the right to use regional facilities for low-level radioactive waste generated outside the central Midwest region of the United States, for the right to use facilities outside the region for waste generated within the region, and to approve the disposal of low-level radioactive waste generated within the region at other than a regional low-level radioactive waste facility.

Activities: To appear before any court of law or any federal, state, or local agency, board, or commission in matters relating to low-level radioactive wastes.

Publications:
Annual report published by commission and distributed by the secretary-treasurer

DOE/NSF Nuclear Science Advisory Committee (Nuclear Research)
Physics Division
Directorate for Mathematical and Physical Sciences
National Science Foundation
1800 G Street, NW
Washington, DC 20550

Established: Committee was originally established September 23, 1977, as the ERDA/NSF Nuclear Sciences Advisory Committee under the Energy Research and Development Administration; name changed in 1978 and transferred to the Department of Energy (which replaced ERDA); transferred to the National Science Foundation, August 23, 1983. Membership: Committee comprises 16 representatives of the scientific community in basic nuclear research appointed by the director of the NSF. Meetings: About 5 times a year.

Purpose: Public advisory committee under the Directorate for Mathematical and Physical Sciences, National Science Foundation, to advise DOE/NSF on scientific priorities of basic nuclear research.

Activities: Encompasses experimental and theoretical investigations of the fundamental interactions, properties, and structures of atomic media. Committee assesses and makes recommendations concerning adequacy of present facilities, facility and instrumentation development programs, relationships of basic nuclear research with other fields of service, and any specialized studies requested by DOE or NSF.

Publications: None

Diagnostic Radiology Study Section
Division of Research Grants
National Institutes of Health
Westwood Building, Room 219
Bethesda, MD 20892

Established: Section was originally established February 12, 1979, by authority of 42 USC 217a, the Public Health Service Act, as amended, as the Diagnostic Radiology and Nuclear Medium study section; name changed in 1980. Membership: Consists of 20 members appointed by the director of the National Institutes of Health for overlapping 4-year terms. Members have knowledge in areas related to the field of diagnostic radiology. Meetings: Section meets 3 times a year, October, February, June.

Purpose: Advise on radiology studies.

Activities: Advises the secretary and assistant secretary of health and human services and the director of the National Institutes of Health regarding applications and proposals for grants-in-aid for research and training activities and proposals relating to the generation, detection, and use and development of radioisotopes and radiopharmaceuticals for diagnostic, bioresearch and therapy, and the medical use and bioeffects of nonionizing radiation.

Publications: None

Dose Assessment Advisory Group (DAAG)
Nevada Operations Office
Department of Energy
P.O. Box 14100
Las Vegas, NV 89114

Established: 1980 by the Department of Energy Active Memorandum. Membership: Group consists of 15 members from scientific fields appointed by the secretary of energy.

Purpose: Advise and make recommendations on radiation doses, including planning, organization, and technical direction.

Activities: Project concerns the evaluation and assessment of the amount of radiation received by members of the off-site population surrounding the Nevada Test Site as a result of the nuclear test operations conducted there.

Publications: None

Environmental Radiation Exposure Advisory Committee
Environmental Protection Agency
401 M Street, SW
Washington, DC 20460

Established: 1965 by the chief, Bureau of State Services, Department of Health, Education and Welfare under Section 222 of the Public Health Service Act. Transferred to the Environmental Protection Agency in 1970. Membership: Committee composed of ten members appointed by the administration, Environmental Protection Agency. Terminated 1976.

Purpose: The purpose was to improve national programs of radiological surveillance for the protection of public health.

Activities: The committee provided guidance and expertise in the development of programs needed to estimate public exposure from ionizing and nonionizing environmental radiation and advised on programs needed to minimize public exposure from environmental radioactivity.

Publications: None

Midwest Interstate Low-Level Radioactive Waste Commission
(Nuclear Waste)
Room 588
350 North Robert Street
St. Paul, MN 55101

Established: Commission was established under authority of the Midwest Interstate Low-Level Radioactive Waste Management Compact, Section 225 of Title II of P.L. 99-240, the Omnibus Low-Level Radioactive Waste Interstate Compact Consent Act of January 15, 1986. Membership: One member from each state—Indiana, Iowa, Michigan, Minnesota, Missouri, Ohio, Wisconsin. Meetings: Commission meets at least once a year.

Purpose: The commission was established to enter into an agreement for the disposal of radioactive waste generated outside the region, for waste generated within the region.

Activities: To appear before any court of law or federal, state, or local agency board or commission in problems related to low-level waste disposal management.

Publications: None

Northeast Interstate Low-Level Radioactive Waste Commission
(Nuclear Waste)
55 Princeton Hightstown Road
Princeton Junction, NJ 08550

Established: Commission was established under the authority of the Northwest Interstate Compact on Low-Level Radioactive Waste Management, Section 221, Title II of P.L. 99-240, the Omnibus Low-Level Radioactive Waste Interstate Compact Consent Act of January 15, 1986. Membership: Commission consists of one member from Connecticut, New Jersey, Delaware, and Maryland. Meetings: Commission meets irregularly.

Purpose: Commission develops, adopts, and maintains a regional management plan to ensure safe and effective management of low-level radioactive waste within the northeastern United States pursuant to the Northeast Interstate Low-Level Radioactive Waste Management Compact. The Commission may enter into agreement with any person, state, regional body, or group of states for the importation of low-level radioactive waste into a region and for the right of access to facilities outside the region for waste generated within the region. It may also grant a generator of low-level waste in the region permission to export the waste to another region.

Activities: Appears as an intervenor in party of interest before any court of law or agency that has jurisdiction of waste disposal.

Publications: None

Northwest Low-Level Waste Compact Committee (Nuclear Waste)
Low-Level Radioactive Waste Program
Office of Nuclear Waste
Washington Department of Ecology
Mail Stop PV-11
Olympia, WA 98504

Established: 1986 under authority of the Northwest Interstate Compact on Low-Level Radioactive Waste Management, Section 211 of Title II of P.L. 99-240, the Omnibus Low-Level Radioactive Waste Interstate Compact Consent Act. Membership: Committee comprises one member, appointed by the governor, from Alaska, Hawaii, Idaho, Montana, Oregon, Washington, and Utah. Committee meets irregularly.

Purpose: Committee has the responsibility for providing protection of citizens from the effects of low-level radioactive wastes and the maintenance of a viable economy in disposal regions.

Activities: Considers problems that arise under the jurisdiction of the compact.

Publications: None

Nuclear Weapons Council
Office of the Secretary
Department of Defense

The Pentagon Room 3E 1074
Washington, DC 20301-3050

Established: Council was established under authority of P.L. 99-661, 1986, to comply with a recommendation by the President's Blue Ribbon Task Group on Nuclear Weapons Management. Membership: Council consists of the director of defense research and engineering, vice chairperson of the Joint Chiefs of Staff, and a senior representative of the Department of Energy appointed by the secretary of energy.

Purpose: An interagency advisory council of the Office of the Secretary of the Department of Defense.

Activities: Council prepares the annual Nuclear Weapons Stockpile Memorandum; develops nuclear weapons options and cost of those options; coordinates nuclear weapons programs between Departments of Defense and Energy. Council also considers safety, security, and controls of existing weapons and proposed new weapons and provides guidance regarding priorities for research on nuclear weapons.

Publications: None

Rocky Mountain Low-Level Radioactive Waste Board (Nuclear Waste)
Suite 100
1600 Stout Street
Denver, CO 80202

Established: Board was established in 1986 under authority of the Rocky Mountain Low-Level Radioactive Waste Compact, Section 226 of Title II of P.L. 99-240, the Omnibus Low-Level Radioactive Waste Interstate Compact Consent Act.

Purpose: Board was established to make information available to participating states and to the general public on low-level waste. Information is gathered on technologies and an inventory is kept of all generators within its regions. Suggestions are made to appropriate authorities. It surveys the needs for regional facilities and capacity to manage them. A regional development plan for burial of low-level waste is part of its duties, as well as an emergency evacuation plan.

Activities: Develops plans to deal with any emergency.

Publications: None

Safeguards and Security Steering Committee (Nuclear Safety)
Department of Energy
Forrestal Building
1000 Independence Avenue, SW
Washington, DC 20585

Established: Committee was established in 1983 at the direction of the National Security Council. Membership: Committee comprises representatives of the Departments of Energy and Defense, the Federal Bureau of Investigation, and the National Security Council.

Purpose: An interagency advisory committee that functions under the Department of Defense to oversee plans and activities designed to improve nuclear safeguards and security.

Activities: Reviews individual site security plans and additional measures to upgrade security protection.

Publications: None

Scientific Advisory Group on Effects (SAGE) (Nuclear Weapons)
Defense Nuclear Agency
Washington, DC 20305

Established: 1965 as a public advisory group to advise the director, Defense Nuclear Agency, Department of Defense. Membership: Consists of 16 members with knowledge and experience in scientific fields associated with nuclear weapons tests and research. At least one person must be a government employee. Meetings held semiannually or on call by the chairman.

Purpose: The major responsibility of the group is to advise the director, Defense Nuclear Agency, on matters relating to nuclear weapons tests and development.

Activities: Group reviews and evaluates long-range plans for the development of nuclear weapons. New approaches and techniques are recommended for determining the effects of nuclear weapons. It also renders advisory assistance to the solution of specific nuclear weapons problems.

Publications: None

Southeast Interstate Low-Level Radioactive Waste Management Commission (Nuclear Waste)
Suite 100-B
3901 Barrett Drive
Raleigh, NC 27609

Established: Commission was established under authority of the Southeast Interstate Low-Level Radioactive Waste Management Compact, Sections 223 of Title II of P.L. 99-240, the Omnibus Low-Level Radioactive Waste Interstate Compact Consent Act of 1986. Membership: Commission consists of two members each from Alabama, Florida, Georgia, Mississippi, North Carolina, South Carolina, Tennessee, and Virginia.

Purpose: Commission is to develop and use procedures for determining, consistent with public health and safety, the type and number of regional facilities necessary to manage low-level radioactive waste generated in the southeastern United States. It has the right to enter into agreements with persons or groups of states for the disposal of radioactive wastes both inside and outside the region.

Activities: Acts as an intervenor before Congress, state legislatures, courts, or agencies over the management of low-level radioactive waste.

Publications: None

Strategic Defense Initiative (SDI)
Strategic Defense Initiative Organization
Department of Defense
The Pentagon
Washington, DC 20301

Established: Founded in 1983 by a directive of President Ronald Reagan. Functions under the Strategic Defense Initiative Organization.

Purpose: To determine the feasibility of eliminating the threat posed by nuclear ballistic missiles and increasing the contribution of defensive systems to U.S. and allied security.

Activities: Conducts research on technologies that could lead to effective defense against ballistic missiles, including space-based defensive systems that could destroy ballistic missiles launched against the United States by a hostile nation. Defense system, SDI, popularly known as "Star Wars."

Publications: None

Strategic Defense Initiative (SDI) Advisory Committee
Strategic Defense Initiative Organization
Department of Defense
The Pentagon
Washington, DC 20301

Established: Founded December 27, 1984, by directives of President Ronald Reagan. Membership: 12 to 24 members who are knowledgeable in technical areas relating to SDI. Members are appointed by the secretary of defense upon recommendation of the director, Strategic Defense Initiative Organization. Meetings: 6 times a year for 2-day sessions.

Purpose: Provides advice to the secretary of defense on matters relating to the Department of Strategic Defense Initiative.

Activities: Conducts research on technologies that could lead to effective defenses against nuclear ballistic missiles and determines the feasibility of eliminating the threat posed by such missiles. Committee's advice includes technical content, program emphasis, and schedule of SDI research and technology programs.

Publications: None

International Organizations

Agency for the Prohibition of Nuclear Weapons in Latin America (OPANAL) (Disarmament)
Temistocles 78

Polanco
Mexico City 011560, D.F. Mexico

Established: 1969. Members: 23. Staff: 7. Biennial conference.

Purpose: To administer the Treaty of Tlatelolco for the Prohibition of Nuclear Weapons in Latin America and promote its goals to establish procedures for an international control system to ensure observance of the treaty and to encourage member countries to ensure exclusively peaceful use of nuclear materials.

Activities: Maintains a library and prepares material for general conference resolutions and speakers.

Publications:
Conference resolutions
Studies

Association of European Atomic Forums (FORATOM) (Nuclear)
Forum Atomique European—FORATOM
1 St. Alban's Street
London SW1Y 45 L
England

Established: 1960. Members: 14. Staff: 6. Meetings: Annual. Languages: English, French, and German.

Purpose: National forums representing power utilities, manufacturers of nuclear plants and components, electronics companies, research institutions, financing bodies, and engineering consultancies from western European countries.

Activities: Seeks the economic development and peaceful uses of nuclear energy. Identifies and studies problems associated with the nuclear industry and proposes solutions and presents opinions to European governments. Promotes better public understanding of issues related to the development and use of nuclear power in the generation of electricity. Maintains relations with the International Atomic Energy Agency.

Publications:
Foratom, annual, status report on nuclear energy developments
Foratom Directory, periodic
Directory of Press Contact, irregular

European Nuclear Disarmament (END)
11 Goodwin Street
London N4 3HQ
England

Established: 1980. Staff: 4. Language: English

Purpose: Promotes peace, international solidarity, and freedom of speech.

Activities: Goal is to make Europe nuclear free, nonaligned, and independent of superpowers. Maintains such specialty working groups as churches, higher education, parliament, press, and women's groups.

Publications:
END Journal, bimonthly, English
END Register, quarterly
END Newsletter, quarterly, English
Special requests

European Nuclear Medical Society (ENMS) (Nuclear Medicine)
c/o Professor Csernay
Medizinische Universitat Szeged
Institut fur Nuklearmedizin
Koranyl Fasor 13
H-6720
Szeged, Hungary

Established: 1974. Membership: 600. Language: English. Meetings: Annual, European Nuclear Medicine Congress.

Purpose: To promote nuclear medicine.

Activities: Professionally active in the science and technology of medical treatment employing radioactivity.

Publications: None

European Nuclear Society (ENS)
(Europaische Kernenergie-Gesellschaft-ENS)
Postfach 2613
CH - 3001
Berne, Switzerland

Established: 1975. Represents 17 national nuclear energy societies with 16,000 members, including European and non-European industrial plants, utility companies, and research institutions. Quadrennial conferences.

Purpose: Promotes the interest of European nuclear energy producers and strives for scientific and engineering progress as it affects the peaceful use of nuclear energy.

Activities: Functions include keeping members informed about current developments in the nuclear energy industry worldwide and representing the industry before European and international bodies. Encourages cooperation among members. Provides scholarships, fellowships, and student exchanges. Conferences are also organized.

Publications:
Nuclear Europe, periodic (English)
Nuclear Newsletter for Euro-Politicians, periodic

European Organization for Nuclear Research (CERN)
(Organisation Europeenne pour la Recherche Nucleaire-CERN)
CH-1211
Geneva 23, Switzerland

Established: 1954. Membership: 14.

Purpose: Goals are to allow for collaboration among physicists and to provide physicists with particle physics (subnuclear or high-energy physics) research facilities unavailable in member countries.

Activities: Organization conducts research primarily for scientific goals and is not concerned with the development of nuclear power systems or weapons. Research is carried out at the Laboratory for Particle Physics in Geneva, Switzerland.

Publications:
CERN Courier, 10/year (English and French)
Annual report

European Standards on Nuclear Electronics Committee (ESONE)
c/o CERN
CH-1211
Geneva 23, Switzerland

Established: 1954. Membership: 100. Annual meeting.

Purpose: The fundamental purpose is to work toward standardization within nuclear electronics.

Activities: This organization encourages work on nuclear electronics in universities and commercial laboratories. A goal of nuclear electronics, a form of electronics that utilizes a nuclear powered battery, is to produce a commercial battery for everyday use.

Publications: None

International Association of Radiopharmacology (IAR)
c/o Mervyn W. Billinghurst
Radiopharmacy, Health Science Centre
700 William Avenue
Winnipeg, MB, Canada R3E 0Z3

Established: 1980. Members: 485. Regional Groups: 15. Meetings: Biennial.

Purpose: To promote information regarding radiotracers used for biological and medical purposes.

Activities: Sponsors teaching programs in medical schools for scientists who utilize radiotracers in biology and medicine. Grants awards for best research works presented during meetings.

Publications:
Newsletter, three per year
Book on radiopharmacology
Plans to publish international journal

International Atomic Energy Agency (IAEA) (Nuclear)
(Agence Internationale de l'Energie Atomique)
Vienna International Centre
Wagramerstrasse 5
Postfach 100
A-1400 Vienna, Austria

Established: 1957. Membership: 113. Staff: 1,861. Holds semiannual meetings.

Purpose: Formed to accelerate and enlarge the contributions of atomic energy to peace, health, and prosperity throughout the world. Ensures that any assistance given by the agency is not used for military applications.

Activity: Encourages research and development of the practical applications of atomic energy for peaceful uses throughout the world and fosters the exchange of scientific and technical information among nations. Has established health and safety standards and applies safeguards in accordance with the Treaty on the Non-Proliferation of Nuclear Weapons.

Publications:
Atomindex, semimonthly
Nuclear Fusion, monthly
Bulletin, quarterly
Also publishes proceedings of conferences, symposiums, and seminars, technical directories, and safety and legal reports.

International Nuclear and Energy Association (IN&EA)
645 N. Michigan Avenue
Suite 860
Chicago, IL 60611

Established: 1981. Membership: 277. Annual meeting.

Purpose: Seeks to develop new sources of energy and coordinates this energy source with existing sources.

Activities: Engineers, scientists, and technical personnel conduct workshops, seminars, and aid research programs. A library is being developed on energy sources.

Publications: None

International Nuclear Law Association (INLA)
29, Square de Meeus
B-1040 Brussels, Belgium

Established: 1970. Members: 451. Budget: $132,500. Biennial congress.

Purpose: To arrange for and promote the study of legal problems related to the peaceful utilization of nuclear energy.

Activities: The association advocates the protection of the human race and the environment and cooperates on a scientific basis with associations and institutions with similar interests. Organizes conferences, discussions, lectures, and seminars.

Publications: None

Nordic Liaison Committee for Atomic Energy (NKA) (Nuclear)
(Nordisk Kontaktorgan for Atomenergisporgsmal - NKA)
Postboks 49
DK - 4000
Roskilde, Denmark

Established: 1957. Membership: 3.

Purpose: Reports on nuclear activities to authorities in Denmark, Finland, Iceland, Norway, and Sweden.

Activities: Coordinates activities related to the peaceful use of nuclear energy. Special concern is given to Nordic nuclear safety.

Publications:
Reports on Nordic nuclear safety projects, periodic.

OECD Nuclear Energy Agency (NEA)
Agence de l'OCDE pour Energie Nucleaire (AEN)
38, Boulevard Suchet
F - 75016
Paris, France

Established: 1958. Membership: 23. Staff: 84. Irregular conferences.

Purposes: The goal of the agency is to promote cooperation in the safety and regulation of nuclear power among OECD countries and to develop nuclear energy as a contribution to economic progress.

Activities: Encourages the exchange of scientific and technical information and the coordination of government regulatory policies and practices. Coordinates and supports research and development programs. Reviews technical and economic aspects of the nuclear fuel cycle. Assesses supply and demand for nuclear power. Areas of concern include safety, radiation protection, waste management, development, general development, legal affairs, and science. Functions through such committees as Nuclear Data, Reactor Physics, Radioactive Waste Management, Nuclear Science, Radiation Protection, and Safety of Nuclear Installations.

Publications:
NEA Newsletter, semiannual
Nuclear Law Bulletin, semiannual
NEA Activity Report, annual
Summary of Nuclear Power and Fuel Cycle Data, annual
Red Book: Uranium Resources, Production and Demand, biennial
Yellow Book: Nuclear Energy and Its Fuel Cycle, quadrennial
Other reports on scientific, economic, technical, and regulatory issues

Organisations des Producteurs d' Energie Nucleaire (OPEN) (Nuclear)
52 Rue de Londres
F-75008
Paris, France

Established: 1973. Membership: 21. Staff: 4.

Purpose: Coordination of European utility companies.

Activities: Provides reports on developments, electric power demand, and problems to European electric producers.

Publications: None

Pacific Concerns Resource Center (PCRC)
c/o Belau Pacific Center
Box 176
Koror, Belau 96940

Established: 1980. Members: 800. Staff: 3. Regional groups: 7. Triennial meetings.

Purpose: To provide information to individuals and groups advocating a nuclear-free Pacific region.

Activities: Coordinates actions of groups in the Pacific area, develops educational resources and supplies and disseminates information and maintains a news clipping service.

Publications:
Bulletin, bimonthly
Directory, annual
Journal, periodic
Special publications

World Information Service on Energy (WISE) (Nuclear)
Postbus 5627
NL-1007 AP Amsterdam
The Netherlands

Established: 1978. Annual meetings.

Purpose: To provide information for antinuclear and safe-energy groups worldwide.

Activities: Disseminates information on nuclear power, nuclear weapon proliferation, energy costs, alternative energy sources, and energy development in the Third World. Maintains a small library.

Publications:
WISE News Communique (in Dutch, English, French, and Spanish), biweekly
The Nuclear Fix: A Guide to Nuclear Activities in the Third World (book)

Sources for Chapter 4

Encyclopedia of Associations, Vol. 1, *National Organizations of the U.S.* Detroit, MI: Gale Research Co., 1988.

Encyclopedia of Associations, Vol. 4, *International Organizations.* Detroit, MI: Gale Research Co., 1988.

Encyclopedia of Governmental Advisory Organizations, 1988–1989, 6th ed. Detroit, MI: Gale Research Co., 1987.

The United States Government Manual, 1987/88. Washington, DC: U.S. Government Printing Office, 1987.

5

Bibliography

MODERN STUDIES OF RADIATION began when Mme. Curie isolated pure radium in 1892. There has been a massive increase since the construction of the atomic bomb in the 1940s and the development of nuclear energy in the 1950s. The literature varies from scientific articles to politically oriented papers calling for the abolition of all nuclear activities. This bibliography provides a wide perspective on the environmental hazards of the nuclear activities.

Reference Works

Burns, Grant. **The Atomic Papers: A Citizen's Guide to Selected Books and Articles on the Bomb, the Arms Race, Nuclear Power, the Peace Movement and Related Issues.** Metuchen, NJ: Scarecrow Press, 1984. 309p. ISBN 0-8108-1692-x.

This annotated bibliography cites books and journal articles dealing with subjects ranging from the invention of the bomb and strategies for its use to nuclear proliferation, arms control, the peace movement, atomic scientists and spies to the art of fission. Popular journals such as *Time, New Republic,* and *Harpers* and government documents, for the most part, have been omitted.

Colen, Donald J. **The ABCs of Armageddon: The Language of the Nuclear Age.** New York: World Almanac/Random House, 1988. 208p. ISBN 0-345-35224-6.

This is a volume of 250 entries of acronyms, euphemisms, and defense establishment nomenclature. The descriptions are particularly concerned with the political and psychological power of the nuclear language. The material reveals the awesome power and danger to civilization of the atomic bomb.

Educators for Social Responsiblity. **Knowing the Nuclear World: A Bibliography of Selected Curricula and Teaching Materials.** Cambridge, MA: Educators for Social Responsibility, 1986. 29p. No ISBN.

An annotated bibliography to be used in the study and teaching of nuclear warfare and nuclear disarmament. Materials cover junior high school through college.

Fleisher, Paul, **Understanding the Vocabulary of the Nuclear Arms Race.** Minneapolis, MN: Dillon Press, 1988. 192p. ISBN 0-87518-352-2.

Publication contains more than 300 descriptive entries. Illustrations are most useful. An appendix lists organizations that want to end the nuclear arms race and also lists commercial power plants. The book, intended for the fifth grade up, contains a bibliography for additional readings.

Gay, William, and Michael Pearson. **The Nuclear Arms Race.** The Last Quarter Century Series, no. 1. Chicago, IL: American Library Association, 1987. 289p. ISBN 0-8389-0467-x.

Bibliography is divided into three parts. Part 1 gives books and articles on the historical background. Part 2 discusses the probability and consequences for nuclear war and military and socioeconomic consequences of the arms race; contains list of references. Part 3 discusses the debate over deterrence and strategic theory and future alternatives of living with nuclear weapons, without nuclear weapons and without war; accompanied by many references.

Guenther, Nancy Anderman. **Children and the Threat of Nuclear War: An Annotated Bibliography.** CompuBibs Series, no. 7. Brooklyn, NY: CompuBibs, 1985. 42p. ISBN 0-914791-06-0.

Annotated bibliography dealing with the psychological aspects of nuclear warfare. Offers literature to help educate children about the threat of nuclear warfare.

Hassler, Peggy M. **Three Mile Island: A Reader's Guide to Selected Government Publications and Government-Sponsored Research Publications.** Metuchen, NJ: Scarecrow Press, 1988. 214p. ISBN 0-8108-2118-4.

This bibliography consists of Pennsylvania government publications—state and county—and also federal government publications dealing with the incident at Three Mile Island. Includes technical reports, government hearings, and journal articles. A glossary and subject and author index are included.

Hawley, Norma J. **Radioactive Waste Management in Canada.** Pinawa, Manitoba: AECL, 1980. 42p. No ISBN.

A bibliography dealing with radioactive waste disposal in Canada.

Lau, Foo-Sun. **A Dictionary of Nuclear Power and Waste Management with Abbreviations and Acronyms.** Research Studies in Nuclear Technology. New York: John Wiley, 1987. 396p. ISBN 0-471-91517-3.

Dictionary includes technological, medical, and geological terms that will help standardize terms used in the field of nuclear power and waste management. Some popular terms such as "not in my backyard" (NIMBY) are included.

Mawson, Colin Ashley. **Processing of Radioactive Wastes.** Review Series (International Atomic Energy Agency), no. 18. Vienna: International Atomic Energy Agency, 1961. 44p. No ISBN.

This bibliography gives references useful in examining the problem of disposal of radioactive waste.

Moe, Christine. **The Environmental Impact of Operations at Los Alamos Scientific Laboratory (LASL).** Public Administration Series-Bibliography, P-767. Monticello, IL: Vance Bibliographies, 1981. 21p. No ISBN.

This bibliography lists books, journals, articles, and government publications dealing with the disposal of radioactive waste at the Los Alamos Scientific Laboratory.

Schlachter, Gail, and Pamela R. Byrne, eds. **Nuclear America: A Historical Bibliography.** ABC-CLIO Research Guides, no. 3. Santa Barbara, CA: ABC-CLIO Information Services, 1984. 184p. ISBN 0-87436-360-8.

This annotated bibliography covers the period 1973 to 1982. The book is divided into five chapters: the first studies the development and dropping of the atomic bombs on Hiroshima and Nagasaki; the second studies the development of atomic weapons in the United States after 1945; the third deals with the cold war; the fourth studies the efforts to control nuclear arms; and the last chapter studies the development of atomic energy as a source of energy, nuclear proliferation, disposal of nuclear wastes, and safety measures. The book contains a detailed, lengthy subject index and an author index. Items in the bibliography deal with the United States or are related to the United States.

Semler, Eric, James Benjamin, and Adam Gross. **The Language of Nuclear War: An Intelligent Citizen's Dictionary.** New York: Perennial Library-Harper and Row, 1987. 325p. ISBN 0-06-055051-1.

Book includes more than 500 entries in alphabetical order to enable the average citizen to become familiar with terms applying to nuclear war. It includes a glossary of acronyms and abbreviations as well as a bibliography for additional reading.

Starr, Philip, and William Pearman. **Three Mile Island Sourcebook: Annotations of a Disaster.** Garland Reference Library of Social Science, vol. 144. New York: Garland, 1983. 411p. ISBN 0-8240-9184-1.

Book covers the period of the construction of Three Mile Island through the second anniversary of the accident. It is divided into three sections. Section 1 gives a chronology of the news media, including the *Harrisburg Patriot, Lancaster Intelligencer Journal, Middletown Press, New York Times,* and *Newsweek,* that gives both local and national views. Section 2 includes state and federal government documents. Section 3 has annotations of articles, books, and professional papers about TMI. A name and subject index and a glossary of terms are included.

Wood, M. Sandra, and Suzanne M. Shultz, comps. **Three Mile Island: A Selectively Annotated Bibliography.** Bibliographies and Indexes in Science and Technology. Westport, CT: Greenwood Press, 1988. 309p. ISBN 0-313-25573-3.

Bibliography lists literature written about the Three Mile Island accident. It has a subject index, cross-referenced author index, and broad headings to approach material by topic. There is a section on popular literature.

Books

Uranium and Radioactivity

Anno, J. N. **Notes on Radiation Effects on Materials.** New York: Hemisphere Publishing, 1984. 342p. ISBN 0-89116-375-1.

This book, written in a very readable style, studies material effects such as electronic components, reactor fuels, and living and organic organisms. The effects of radiation on steel are discussed at length. The book is useful for both students and professionals.

Australia Department of Foreign Affairs. **Uranium, the Joint Facilities, Disarmament and Peace.** Canberra, Australia: Australian Government Publishing Service, 1984. 33p. ISBN 0-644-03480-7.

Examines whether the export of Australian uranium and the joint United States–Australian defense facilities contribute to, or jeopardize, arms control.

Eisenbud, Merril. **Environmental Radioactivity.** 3d ed. Orlando, FL: Academic Press, 1987. 475p. ISBN 0-12-235153-3.

Much of this third edition is devoted to a discussion of the natural, industrial, and military sources of radioactive contamination of the environment. The chapter on accidental environmental contamination describes events that have caused the release of radioactive material into the environment. The book provides information on the 1979 Three Mile Island accident in the United States and the 1986 accident at Chernobyl in the Soviet Union. This book, an excellent reference work on environmental radioactivity, includes an extensive bibliography and references.

Hall, Eric J. **Radiation and Life.** 2d ed. New York: Pergamon Press, 1984. 255p. ISBN 0-08-028819-7.

The medical use of radiation is accepted and X-ray facilities are available in all hospitals. We receive radiation from natural resources, such as radioactivity in the ground and in food, as well as cosmic rays from outer space. When we fly in commercial jets, we get an extra exposure to cosmic rays. The book has two main purposes. First, to tell what radiation is and how it affects living things; and second, to discuss the sources of radiation to which man is exposed as a result of modern technology. A comparison is made of medical and industrial uses of radiation. Written in nontechnical style aimed at educating the public about the use and effects of radiation, the book contains many pictures, tables, and graphs. The book excludes discussion of nuclear weapons.

Hurley, Patrick M. **Living with Nuclear Radiation.** Ann Arbor, MI: University of Michigan Press, 1982. 131p. ISBN 0-472-09339-8.

This book concentrates solely on nuclear radiation. It aims to provide information about the natural background levels of radioactivity that have been with us for a long time that are harmful to people. Such things as radioactivity in the ground, radiation from common rocks and cosmic rays, doses of natural radioactivity within the body, estimates of risk to human health, and man-made radioactivity are discussed. The book has a glossary and selected references.

Lillie, David W. **Our Radiant World.** Ames, IA: Iowa State University Press, 1986. 226p. ISBN 0-8138-1296-8.

The author begins with a discussion of radiation, explaining what it is. He then points out the various types of radiation—what they are, where they come from, how we can use them. He then discusses how radiation affects us in our daily life. He poses the questions, To what extent should radiation be used in medicine? and What should be done with nuclear weapons and nuclear power? Radiation has been around for billions of years, so we should understand it and control it by using it wisely. Radiation is neither good nor evil. Since the book was written, the Chernobyl incident occurred; thus a chapter on Chernobyl was added at the end. The book has a selected critical bibliography.

Miller, Kenneth L., and William A. Weidner, eds. **Handbook of Management Radiation Protection Programs.** Boca Raton, FL: CRC Press, Inc., 1986. 498p. ISBN 0-8493-3769-0.

This book discusses lessons learned through mistakes, radiation litigations, and the aftermath of Three Mile Island. Of special interest to persons responsible for interstate shipment of radioactive materials, the book is also a valuable source of information for persons responsible for radiation emergency planning.

Neff, Thomas L. **The International Uranium Market.** Cambridge, MA: Ballinger Publishing, 1984. 333p. ISBN 0-88410-850-3.

Part 1 examines the structure of the international market, its history, and supply and demand. Part 2 analyzes the uranium industry in main producer nations and each nation's role in international fuel trade. Part 3 gives an overview of international nuclear fuel trade and explores its future.

Owen, Anthony David. **The Economics of Uranium.** Praeger Special Studies: Praeger Scientific. New York: Praeger Publishers, 1985. 217p. ISBN 0-03-003799-9.

Gives the history of uranium, which was of very little importance until late in the nineteenth century. The economic structure of the uranium industry is analyzed. Such topics as the nuclear fuel cycle, world resources and reserves, production, demand, uranium enrichment, and prospects are discussed. Contains a lengthy bibliography, figures, and tables.

Pochin, Sir Edward. **Nuclear Radiation: Risks and Benefits.** Oxford, England: Clarendon Press, 1983. 197p. ISBN 0-19-858329-X.

Begins by reviewing the history of the discovery of X-rays and the potential danger. It examines the radiation we receive from nature and man-made sources, genetic effects of radiation, effects of high doses, radiation protection from medical and occupational exposures, and the issues of safety and risk.

Reinig, William C., ed. **Environmental Surveillance in the Vicinity of Nuclear Facilities.** Proceedings of a symposium sponsored by the Health Physics Society. Springfield, IL: C. C. Thomas, 1970. 465p. No ISBN.

Engineers and scientists from nine nations participated in the symposium held in Augusta, Georgia, January 24 to 26, 1968. The participants reviewed experiences, accomplishments, and current research in detection and control of environmental radioactivity. They examined the objectives and direction of environmental surveillance in the vicinity of nuclear facilities in the future.

Ringholz, Raye. **Uranium Frenzy: Boom and Bust on the Colorado Plateau.** New York: W. W. Norton, 1989. 310p. ISBN 0-393-02644-2.

After World War II with the testing of the atomic bomb and later the development of nuclear energy, the U.S. government indicated there was a need for increasing the supply of uranium. This demand set the stage for a massive surge of exploration and development by miners in the American Southwest. This book presents a vivid picture of the activities of the companies and the individuals who sought this new resource. The uranium industry has experienced many vicissitudes. Because demand has fluctuated greatly, many companies have failed while other companies have flourished. The tragedy resulting from the lack of protecting miners from exposure to radiation by disregarding the warning of scientists to provide mine ventilation is strikingly presented. This is the story of "boom and bust" of uranium mining in the United States in the past 40 years.

Turner, James E. **Atoms, Radiation and Radiation Protection.** New York: Pergamon Press, 1986. 324p. ISBN 0-08-031937-8.

Properties of various kinds of ionizing radiation and their interactions with matter are discussed. Radioactive sources that produce radiation are described.

United Nations Environmental Programme. **Radiation, Doses, Effects, Risks.** Nairobi, Kenya, 1985. 64p. ISBN 92-807-1104-0.

In recent decades few scientific issues have gained the attention of the general public more than the dangers of radiation. Because of the emotional aspects of this issue, much of the information is inaccurate. This volume, prepared by the United Nations Scientific Committee on the Effects of Atomic Radiation (UNSCEAR), presents evidence on the sources and effects of radiation. A wide range of topics is presented on natural and man-made sources. The conclusions are that high doses cause severe tissue damage and ultimately kill the individual. Low-level doses cause cancer and induce genetic defects. But most important, the greatest sources of radiation for the general public are natural sources. Nuclear power contributes only a small proportion of radiation emitted by human activities. Summarizes the present-day information available on the sources of radiation and the risks of exposure to human beings.

Uranium Institute. **Uranium and Nuclear Energy: 1987. Proceedings of the Twelfth International Symposium Held by the Institute, London, 2–4 September 1987.** London: 1988. 454p. ISBN 0-94677-11-x.

This symposium examined such topics as public acceptance, reactor safety, the market balance, and the impact of trade restrictions on that balance.

U.S. National Council on Radiation Protection and Measurements. **A Handbook of Radioactivity Measurement Procedures.** 2d ed. NCRP Report no. 58. Bethesda, MD: 1985. 592p. ISBN 0-913392-71-5.

Represents an updating of radiation detection techniques. It contains useful information on such matters as detector calibration and source preparation. An excellent reference book on precise measurement of radioactive sources.

Nuclear Warfare

General Works

Ball, Desmond, and others. **Crisis Stability and Nuclear War.** Ithaca, NY: Peace Studies Program, Cornell University, 1987. 105p. No ISBN.

This is a report published under the auspices of the American Academy of Arts and Sciences and the Cornell University Peace Studies Program. Chapter 1 defines the word "crisis." Chapter 2 sketches that setting. Chapter 3 examines the various ways in which superpower crises may occur. Chapter 4 examines generic features of events that might carry a superpower crisis to armed conflict and then to nuclear war. The last chapter summarizes technological trends that may have an impact on crisis stability. The book contains a lengthy glossary.

Day, Samuel H., Jr., ed. **Nuclear Heartland: A Guide to the 1,000 Missile Silos of the United States.** Madison, WI: Progressive Foundation, 1988. 95p. ISBN 0-942046-01-3.

Deals with the Great Plains area locations of underground launch sites and launch control centers. Activities of antinuclear groups are also treated.

de Leon, Peter. **The Altered Strategic Environment: Toward the Year 2000.** Lexington, MA: Lexington Books, 1987. 113p. ISBN 0-669-14576-9.

Discusses in detail the relationships among the factors of a nuclear winter, new capabilities of conventional weapons, the Strategic Defense Initiative, and the opinion of the public.

Gerson, Joseph, ed. **The Deadly Connection: Nuclear War and U.S. Intervention.** Philadelphia, PA: New Society Publishers, 1986. 253p. ISBN 0-86571-068-6.

A conference, "Deadly Connection," was held in Cambridge in December 1982 under the auspices of the New England office of the American Friends Service Committee. This book is a compilation of articles and speeches from that meeting. A study was made to show the relationship between U.S. nuclear policy and U.S. military intervention around the world and how it might be changed. Book starts by defining "deadly connection" and then discusses first-strike policy and U.S. intervention, Third World nuclear triggers, and finally strategies for survival.

International Peace Academy. **The Indian Ocean as a Zone of Peace.** IPA Report no. 24. New York: 1986. 155p. ISBN 0-89838-916-x.

Presents papers delivered at a workshop organized by the Bangladesh Institute of International and Strategic Studies and the International Peace Academy, which was held in Dhaka, Bangladesh, November 23–25, 1985. The workshop was attended by 66 participants and 38 observers from 24 countries. The aim of the workshop was to analyze factors that prevent settlement and enhance conflicts; encourage regional cooperation; discuss ways to diffuse conflict points and choose peaceful resolution to conflicts. Papers covered such topics as the war in Afghanistan, USSR in Afghanistan, and superpowers in the Indian Ocean—their goals and impact of internal ethnic conflicts on the region. A list of participants is given.

Karem, Mahmoud. **A Nuclear-Weapon-Free Zone in the Middle East: Problems and Prospects.** Contributions to Military Studies no. 65. Westport, CT: Greenwood Press, 1988. 186p. ISBN 0-313-25628-4.

Karem is Counsellor with the Ministry of Foreign Affairs of Egypt and a member of the Cabinet of the Deputy Prime Minister and Minister of Foreign Affairs. He also teaches at the American University in Cairo. He suggests the possibility of eliminating nuclear weapons because they pose such great danger to civilization. The book deals with the Middle East and shows efforts made to establish a nuclear-free zone and the problems, prospects, and recommendations encountered.

Kelleher, Catherine M., Frank J. Kerr and George H. Quester, eds. **Nuclear Deterrence: New Risks, New Opportunities.** Washington, DC: Pergamon-Brassey's International Defense Publishers, 1986. 238p. ISBN 0-08-032783-4.

This book was the aftermath of the University of Maryland's International Security Project (MISP) held in September 1984 to discuss good and bad news regarding nuclear war and nuclear deterrence. More than 500 scholars, government officials, and concerned citizens attended the three-day conference at College Park. Book contains papers presented on such topics as nuclear winter, nature of nuclear deterrence, new technologies in nuclear arms, arms control and defense policy, and perspectives on the general nature of global military balance.

Kipnis, Kenneth, and Diana T. Meyers, eds. **Political Realism and International Morality: Ethics in the Nuclear Age.** Boulder, CO: Westview Press, 1987. 271p. ISBN 0-8133-0456-3.

Three themes are discussed in the book. Part 1 examines morality and the international order; Part 2 examines the ethics of nuclear deterrence; and Part 3 examines nationalism and the prospects for peace. The book should be of interest to scholars dealing with international affairs and those who wish to explore the implications of morality and reason in foreign policy.

Lefever, Ernest W. and E. Stephen Hunt, eds. **The Apocalyptic Premise: Nuclear Arms Debated.** Washington, DC: Ethics and Public Policy Center, 1982. 417p. ISBN 0-89633-062-1.

Collection of 31 essays by statesmen, religious leaders, scholars, and journalists on such subjects as arms control issues, the peace movement, the apocalyptic premise, and churches and nuclear arms.

Lewis, William H. **The Prevention of Nuclear War: A United States Approach.** United Nations Institute for Training and Research. Boston, MA: Oelgeschlager, Gunn, and Hain Publishers, 1986. 103p. ISBN 0-89946-206-5.

A UNITAR study dealing with disarmament, international peace and security, and the prevention of a nuclear war.

Mandelbaum, Michael. **The Nuclear Future.** Cornell Studies in Security Affairs. Ithaca, NY: Cornell University Press, 1983. 131p. ISBN 0-8014-9254-8.

The Nuclear Future had its origin as a paper for a conference at the Naval War College in Newport, Rhode Island, and finally expanded into a book. In Chapter 1 the author gives an overview of nuclear energy in the world. Chapter 2 discusses the nuclear arsenals in the United States and the Soviet Union showing the relationship between the superpowers. Chapter 3 examines the spread of nuclear weapons to countries that do not have them. Chapter 4 examines the antinuclear movements that have developed around the world. The last chapter predicts the future of nuclear weapons. Contains a glossary and selected references.

Patterson, Walter C. **The Plutonium Business and the Spread of the Bomb.** Published for the Nuclear Control Institute. San Francisco, CA: Sierra Club Books, 1984. 272p. ISBN 0-87156-837-3.

The main objective of this book is to simplify for the public the evolution of plutonium from atomic bombs to a fuel that may be used in nuclear power plants. Suggestions are made on how to stop the global spread of plutonium before it gets out of control. Plutonium has been around since 1940, being the first artificially produced element made in quantities. There are now over 400,000 kilograms worldwide, about half being separated from radioactive fuel that is now in the world's stockpile of nuclear weapons. Some is being used for research and development and plutonium breeder reactor fuel or waiting to be recycled in nonbreeder nuclear power plants, and some is still unseparated and stored for separation or disposal. Plutonium has a dual nature: It can power cities or it can blow them up. The book is divided into three parts: Part 1, Plutonium Dreams—1946–74; Part 2, Plutonium Diplomacy, 1974–79; and Part 3, Plutonium Addiction: Curable or Terminal?—1980 and After. A bibliography of books and magazines gives the reader additional material to read.

Pitt, David, and Gordon Thompson, eds. **Nuclear Free Zones.** London: Croom Helm Ltd., 1987. 145p. ISBN 0-7099-4076-9.

The book is based on Geneva International Research Institute's research program to study the possibilities for peace. The Institute for Resource and Security Studies is a cosponsor. The book contains papers exploring the theory and practice of nuclear-free zones while searching for peace. Topics include: the Treaty for the Prohibition of Nuclear Weapons in Latin America; the Treaty of Rarotonga: South Pacific Nuclear-Free Zone; nuclear-free Europe; nuclear-free zones in Africa; and problems and prospects of nuclear-free zones.

Schiff, Benjamin N. **International Nuclear Technology Transfer: Dilemmas of Dissemination and Control.** Totowa, NJ: Rowman and Littlefield, 1983. 226p. ISBN 0-86598-139-6.

Discusses the dissemination and control of nuclear technology and the activities of the International Atomic Energy Agency.

Schwartzman, David. **Games of Chicken: Four Decades of U.S. Nuclear Policy.** New York: Praeger Publishers, 1988. 233p. ISBN 0-275-92884-5.

The author begins by discussing two risks, preemption risk (PR) and aggression risk (AR), along with three myths—the myth of the nuclear-strategy expert, the Soviet bogey, and the economic myth. He then dwells on Truman's choice, the economics of Eisenhower, Kennedy and the national resolve, Nixon the negotiator, Carter and the MX, and Reagan the continuer. The author then reviews the discussion of the five issues that were the center of the debate in the last few years—the freeze proposal, the no-use proposal, the no-first-use proposal, the build-down proposal and last, proposals to reduce the preemption risk (PR). The book ends with a list of acronyms and a selected bibliography.

Williams, Robert C., and Philip L. Cantelon, eds. **The American Atom: A Documentary History of Nuclear Policies from the Discovery of Fission to**

the Present, 1939–1984. Philadelphia, PA: University of Pennsylvania Press, 1984. 333p. ISBN 0-8122-7920-4.

Gives a documented history of American atomic development and how it has achieved so much power that it could mean self-destruction yet is considered essential to protect our national security. We benefit from its power to produce electrical energy, but at the same time we must consider its risks.

Antinuclear Movement

Boyle, Francis A. **Defending Civil Resistance under International Law.** Dobbs Ferry, NY: Transnational Publishers, 1987. 378p. ISBN 0-94130-43-X.

Written as a legal manual to be used in preparing the defense of citizens protesting to uphold international law related to the nuclear area. There is some focus on antinuclear protests.

Dewar, John, and others, eds. **Nuclear Weapons, the Peace Movement and the Law.** London: Macmillan Press, 1986. 255p. ISBN 0-333-41410-1.

Examines public lectures organized by the School of Law, University of Warwick, Coventry, England, October 1983–March 1984 and papers presented at the International Conference on the Legality of Nuclear Weapons, Coventry, May 27, 1984.

Dougherty, James E., and Robert L. Pfaltzgraff, Jr. **Shattering Europe's Defense Consensus: The Antinuclear Protest Movement and the Future of NATO.** Washington, DC: Pergamon-Brassey's International Defense Publications, 1985. 226p. ISBN 0-08-032770-2.

Examines strategy, politics, and ethical feelings of the protest movement. The antinuclear movement in Britain, West Germany, the Netherlands, Italy, and France is discussed in detail.

Johnston, Carla B. **Reversing the Nuclear Arms Race.** Cambridge, MA: Schenkman Books, 1986. 194p. ISBN 0-87047-033-7.

Examines the U.S. nuclear arms policy and discusses the effectiveness of public movements in achieving a nuclear freeze.

Katz, Milton S. **Ban the Bomb: A History of SANE, the Committee for a Sane Nuclear Policy, 1957–1985.** Contributions in Political Science Series, no. 147. Westport, CT: Greenwood Press, 1986. 215p. ISBN 0-313-24167-8.

SANE was organized in June 1957 in New York City by prominent citizens who actively participated against nuclear weapons for years. It became the largest and most influential nuclear disarmament organization in the United States. Gives the history of the activities of the organization.

Kavka, Gregory S. **Moral Paradoxes of Nuclear Deterrence.** Cambridge, England: Cambridge University Press, 1987. 243p. ISBN 0-521-33043-2.

The book, written by a person who experienced radiation-caused illness, stresses the importance of avoiding nuclear war and attaining nuclear disarmament. This is a book in applied ethics. In Part I, Chapters 1 and 2 emphasize the fact that utilitarian costs and benefits of making deterrent nuclear threats may be different from the costs and benefits of carrying out the threat if deterrence fails. Chapter 3 emphasizes deterrence, utility, and rational choice. Chapter 4 points out that there is a conflict in evaluating deterrence between plausible deontological principles. Part II argues that nuclear deterrence is justifiable as a lesser evil by discussing alternatives to deterrence.

Laffin, Arthur J., and Anne Montgomery, eds. **Swords into Plowshares: Nonviolent Direct Action for Disarmament.** San Francisco, CA: Harper & Row, 1987. 243p. ISBN 0-06-064911-9.

Discusses the dangers of a nuclear war and the efforts being made to avert the prospect of a global holocaust. Part I shows how the nuclear arms race threatens our lives and why people engage in acts of divine obedience. It includes a chronology of Plowshares actions where people have entered nuclear weapons facilities or military bases and enacted the Biblical prophecy of "beating swords into plowshares." Part II discusses the meaning of certain acts of civil and divine disobedience and how people reacted to the consequences of court and jail. Part III describes nonviolent resistance actions to nuclear war preparations. Parts IV and V include appendices on the nuclear arms race, the immorality of nuclear weapons, and a listing of national peace and disarmament groups and resistance groups. There is also a lengthy bibliography.

Laqueur, Walter, and Robert Hunter, eds. **European Peace Movements and the Future of the Western Alliance.** New Brunswick, NJ: Transaction Books, Rutgers—The State University, 1985. 450p. ISBN 0-88738-035-2.

Published in association with the Center for Strategic and International Studies, Georgetown University. The book concerns the future of the Western alliance and the development of the peace movement, in Europe mainly, but in the United States to a lesser degree.

Loeb, Paul R. **Hope in Hard Times: America's Peace Movement and the Reagan Era.** Lexington, MA: Lexington Books, 1987. 322p. ISBN 0-669-12929-1.

A narrative account of experiences of various people across the United States with the antinuclear movement. It is written in a very readable style and is most interesting. The *Christian Science Monitor* stated "that it is a disturbing lesson: those who are most directly involved in nuclear work are often those who think least about its implications."

London, Herbert I. **Armageddon in the Classroom: An Examination of Nuclear Education.** Lanham, MD: University Press of America, 1987. 127p. ISBN 0-8191-6547-6.

Deals with the study and teaching of nuclear warfare and steps being taken in the antinuclear movement. Book has a lengthy list of curriculum guides and teacher resources.

Arms Control

Miller, Steven E., ed. **The Nuclear Weapons Freeze and Arms Control.** Cambridge, MA: Ballinger Publishing Co., 1984. 204p. ISBN 0-88730-010-3.

Compilation of papers presented at a conference held January 13–15, 1983, at Harvard University, sponsored by the American Academy of Arts and Sciences and the Center for Science and International Affairs. Papers deal with public opinion and the freeze movement, consequences of a freeze, political perspectives, and Soviet interests and initiatives and discuss the question, Where do we go from here?

Seaborg, Glenn T., and Benjamin S. Loeb. **Stemming the Tide: Arms Control in the Johnson Years.** Lexington, MA: Lexington Books, 1987. 495p. ISBN 0-669-13105-9.

Examines the arms control events and negotiations among nations involved during the Johnson administration. The administration was criticized for substantive positions it took on some issues and criticized for some of the processes it employed that were not always orderly. The authors hope we have learned from mistakes made so they will not be repeated. The authors believe that more was not accomplished during the Johnson administration because of inhospitable world conditions and partly because of the president himself. A list of abbreviations and acronyms used in the book is included.

Thompson, Kenneth W., ed. **Gerard Smith on Arms Control.** W. Alton Jones Foundation Series on Arms Control, vol. 1. Lanham, MD: University Press of America, 1987. 270p. ISBN 0-8191-6451-8.

Smith, the chief U.S. delegate to the Strategic Arms Limitation Talks held in Helsinki and Vienna from 1969 to 1972, produced Volume 1 in the series. He has great knowledge of facts of arms control and has worked untiringly for arms agreements. In the book he discusses arms control and national defense, the international law of nuclear arms control, nuclear proliferation, prospects for arms control negotiations, and Reagan's arms control policies.

———. **Sam Nunn on Arms Control.** W. Alton Jones Foundation Series on Arms Control, vol. 5. Lanham, MD: University Press of America, 1987. 327p. ISBN 0-8191-6615-4.

Nunn, the leading authority in the Senate on defense and arms control, expresses his views on arms control in one of the books in a series written by supporters and critics. This book is the fifth in the series. Nunn states that arms control requires knowledge and understanding of the subject because it is so highly technical. Negotiators must also understand how U.S. actions might affect NATO allies and must have the confidence of the people in actions dealing with arms control. Nunn answers questions about combat readiness, the defense budget, and the Defense Reorganization Act of 1986. A large portion is devoted to a discussion of the ABM Treaty.

Effects of Atomic Bombing

Ball, Howard. **Justice Downwind: America's Atomic Testing Program in the 1950s.** New York: Oxford University Press, 1986. 280p. ISBN 0-19-503672-7.

This is the story of the development and testing of the atomic bomb, with the dangers and contradictions it posed and the impact on our social system. The potential destructiveness of the bomb was evident, but scientists continued work on developing and improving it. The Atomic Energy Commission was created to ensure proper safety standards to protect all people, especially those who lived downwind of the atomic testing area. An interesting feature of the book is the quotations of leaders in the field, including Lewis Strauss, Thomas E. Murray, Gordon Dean, Stewart Udall, and others, that appear at the beginning of each chapter to get you thinking before reading the chapter. Appendix A discusses the long journey of *Bulloch v. United States* (1953)—a case stemming from sheep that died of radiation. Appendix B discusses the medical controversy over the association between low levels of radiation and cancer occurrence. A bibliography includes books, journal articles, congressional hearings, court cases, and references to newspaper articles.

Clark, Ian. **Nuclear Past, Nuclear Present: Hiroshima, Nagasaki and Contemporary Strategy.** Westview Special Studies. Boulder, CO: Westview Press, 1985. 146p. ISBN 0-8133-7049-3.

This book dwells on the bombings of Hiroshima and Nagasaki and the lessons we should have learned. The five lessons are: (1) the first use of atomic weapons was governed by a rational process; (2) lack of control existed over the operations; (3) the bombings were decisive in ending the war and responsible for saving lives; (4) the moral lessons are ambivalent; and (5) we need a clearer perception of the relationship between the atomic experience of 1945 and the postwar nuclear regime. The book has a lengthy bibliography.

Dotto, Lydia. **Planet Earth in Jeopardy: Environmental Consequences of Nuclear War.** New York: John Wiley, 1986. 134p. ISBN 0-471-99836-2.

A popular overview of two large, detailed technical volumes—*The Environmental Consequences of Nuclear War: I, Physical and Atmospheric Effects* and *II, Ecological and Agricultural Effects.* Author discusses consequences of nuclear war, smoke and dust, climatic consequences, changes in the chemistry of the atmosphere, radiation and fallout, agriculture and ecosystems after nuclear war, and human response. Written for the lay reader.

Forsyth, Christopher. **Can Australia Survive World War III?** New York: Rigby Publishers, 1984. 205p. ISBN 0-7270-1877-9.

Forsyth, a former newspaper columnist and editor turned free-lance writer, has produced a very readable book. It deals with Australia and endeavors to make people aware of the strength of the world's supply of nuclear weapons and the devastation they could cause if unleashed. The author delves into the effects of radiation, plans for survival if the bomb is dropped, and what life might be like after a nuclear war.

Fuller, John G. **The Day We Bombed Utah: America's Most Lethal Secret.** New York: New American Library, 1986. 268p. ISBN 0-453-00457-1.

Shows results of a series of atomic bomb tests conducted by the Atomic Energy Commission in the early 1950s in southwestern Utah and eastern Nevada.

Greene, Owen, Ian Percival, and Irene Ridge. **Nuclear Winter: The Evidence and the Risks.** New York: B. Blackwell, 1985. 216p. ISBN 0-7456-0176-6.

Smoke is the main topic of this book. Smoke from fires started by the heat of nuclear fireballs could blot out nearly all of the sunlight from half the Earth for weeks on end; summer could be turned into winter very fast. In a large-scale nuclear war, many plant and animal species could become extinct, and extinction of the human race would be possible. The book concludes with a chapter on nuclear winter—a threat and a challenge. Nuclear winter threatens everyone's future, so ways must be found to remove that threat to our existence. Ten appendices give additional information.

Grinspoon, Lester, ed. **The Long Darkness: Psychological and Moral Perspectives on Nuclear Winter.** New Haven, CT: Yale University Press, 1986. 213p. ISBN 0-300-03663-9.

Symposium papers presented at the 1983 annual meeting of the American Psychiatric Association held in Los Angeles. Carl Sagan talked about policy implications of nuclear war and climatic catastrophe. John E. Mack discussed the reconsideration of national security. Other topics dealt with opposing the nuclear threat, evolutionary and developmental considerations, nuclear winter, and the will to power. Most of the chapters provide references for additional information.

Hacker, Barton C. **The Dragon's Tail: Radiation Protection in the Manhattan Project, 1942–1946.** Berkeley, CA: University of California Press, 1987. 258p. ISBN 0-520-05852-6.

The author, a historian, under the sponsorship of the Reynolds Electrical and Engineering Company and the U.S. Department of Energy undertook the task of documenting radiation protection measures relating to the nuclear weapons test program. The book is more historical than technical in describing the suspense and worries about radiation protection. The Manhattan Project, Metallurgical Laboratory, and Los Alamos are given special attention in developing radiation protection. The Hanford Works is merely mentioned.

Katz, Arthur M. **Life after Nuclear War: The Economic and Social Impacts of Nuclear Attacks on the United States.** Cambridge, MA: Ballinger Publishing Co., 1982. 422p. ISBN 0-88410-096-0.

The author was a member of the Arms Control Seminar at MIT, which studied the effects of nuclear war on Massachusetts. The study expanded to include the nation. From this seminar the report *Economic and Social Effects of Nuclear War on the United States,* written for the Joint Committee on Defense Production of the U.S. Congress, evolved. Much of this book is based on that report. The purpose of the book is to be sure everyone knows the meaning of nuclear war in all its aspects.

Liebow, Averill A. **Encounter with Disaster: A Medical Diary of Hiroshima, 1945.** New York: W.W. Norton, 1970. 209p. ISBN 0-393-05421-7.

This diary discusses the formation of the Joint Commission for the Investigation of the Effects of the Atomic Bomb in Japan and the working relationship with Japanese medical investigators after the hostilities ceased. Much of the material for the book was furnished by the Armed Forces Institute of Pathology. Photographs were taken by the Banka-Sha Agency at Hiroshima as well as by the author himself. Men of the Yale Medical School were participants in the project. Dr. Liebow paints a grim picture of the aftermath of Hiroshima.

London, Julius, and Gilbert F. White. **The Environmental Effects of Nuclear War.** AAAS Selected Symposium 98. Boulder, CO: Westview Press, 1984. 203p. ISBN 0-8133-7014-0.

Assesses information regarding major scientific problems related to the environment that might occur in case of a nuclear war. Some of the topics examined are radiation effects on people; nuclear explosions and atmospheric ozone; long-term ecological effects of nuclear war; transport and residence times of airborne radioactivity; environmental effects; and unresolved problems in the evaluation of environmental effects of nuclear war.

The final chapter discusses what the scientist can do to help solve the problems.

Miller, Richard L. **Under the Cloud: The Decades of Nuclear Testing.** New York: The Free Press, 1986. 547p. ISBN 0-02-921620-6.

This volume presents the development of nuclear testings from the discovery of radium by the Curies about 1900 to the present day. In chronological order such events are discussed as the Hiroshima and Nagasaki explosions, the military testing during the 1950s, and the problems of nuclear testing. A world approach is presented with both the American and Russian developments discussed. A great deal of information is presented. To provide background to nuclear developments, the author provides a setting for the events of the period. The events of nuclear testing are given, but little analysis of their importance is provided.

Neal, James V., and W. J. Schull. **The Effect of Exposure to the Atom Bombs on Pregnancy Termination in Hiroshima and Nagasaki.** National Research Council, Publication 461. Washington, DC: Atomic Bomb Casualty Commission, Hiroshima, Japan, 1956. 241p. No ISBN.

This monograph is a report on efforts during 1946–1955 to answer questions such as: Can any differences between children whose parents were exposed to the atom bomb and those whose parents were not exposed be observed, and if so, what are the differences? The book considers such aspects as the sex ratio, malformation data, stillbirth data, birthweight data, and death following delivery. Many tables and figures are presented throughout the book.

Pittock, A.B., and others. **Environmental Consequences of Nuclear War.** SCOPE Series, 28. New York: John Wiley, 1985. 2 vols. ISBN 0-471-90898-3.

In 1982 more than 300 scientists from over 30 countries joined in an effort to study the possible environmental consequences of nuclear war. They examined the effects on people and their culture and on life support systems of water, air, and soil. Three conclusions they made were that nuclear detonations (1) would cause physical effects from blast, thermal radiation, and local fallout, (2) would cause climatic as well as societal perturbations, and (3) would disrupt communications and power.

Denuclearization

Aizenstat, A. J. **Survival for All: The Alternative to Nuclear War with a Practical Plan for Total Denuclearization.** New York: Billner and Rouse, 1985. 216p. ISBN 0-932755-14-3.

Technology plays a major role in the production of nuclear weapons and the direction taken by the nuclear arms race, but its existence is a political issue. Political action must be taken to solve the challenge posed by nuclear weapons. The second part of the book deals with denuclearization, with one chapter giving a denuclearization plan to avoid a nuclear war.

Byers, R. B., ed. **The Denuclearization of the Oceans.** New York: St. Martin's Press, 1986. 270p. ISBN 0-7099-3936-1.

Based on papers presented at the Conference on the Denuclearization of the Oceans, held in Norrtälje, Sweden, May 1984, and sponsored by the Myrdal Foundation, Sweden, and the International Ocean Institute. Discusses the use of oceans by nuclear-powered naval vessels and naval vessels carrying nuclear weapons. A nuclear weapons chronology covering 1946 through 1985 is given in an appendix. For denuclearization, both as process and as product, to become politically salient, the United States and the Soviet Union must agree that mutual security, rather than unilateral security, is the path to enhance international peace.

Nuclear Nonproliferation

Beckman, Robert L. **Nuclear Non-Proliferation: Congress and the Control of Peaceful Nuclear Activities.** A Westview Special Study. Boulder, CO: Westview Press, 1985. 446p. ISBN 0-8133-7040-x.

On March 10, 1978, President Jimmy Carter signed the Nuclear Non-Proliferation Act of 1978 to limit the further spread of nuclear weapons and materials and facilities needed to make them. Part 1 deals with the era of secrecy (1945–1954) and the race to cooperative competition (1954–1970). Part 2 examines the hazards. Part 3 explains the legislative maneuvering. Part 4 examines congressional constraints on presidential flexibility, and the last part is a reexamination of the act by Reagan and Congress and international responses to it.

Brenner, Michael J. **Nuclear Power and Non-Proliferation: The Remaking of U.S. Policy.** Cambridge, England: Cambridge University Press, 1981. 324p. ISBN 0-521-23517-0.

The book begins by giving the history of the postwar nuclear era. The military and civilian roles of atomic power are discussed. Three crucial episodes in the making of U.S. policy on nuclear exports and weapons proliferation are examined. Chapter 2 deals with the crisis of nuclear enrichment services. Chapter 3 explains the Ford administration's rethinking of established programs and attitudes. Chapters 4 and 5 deal with Carter's nuclear policy and how Carter's administration fared. The last chapter assesses U.S. performance in trying to restore control over the growth and extension of nuclear power worldwide.

Dean, Jonathan. **Watershed in Europe: Dismantling the East-West Military Confrontation.** Lexington, MA: Lexington Books, 1987. 286p. ISBN 0-669-11120-1.

Discusses the future of the NATO–Warsaw Pact confrontation over the next twenty years. The outlook for arms limitation and nuclear nonproliferation is examined.

Dewitt, David B., ed. **Nuclear Non-Proliferation and Global Security.** New York: St. Martin's Press, 1987. 283p. ISBN 0-312-00367-6.

Consists of papers presented at the Conference on Global Security and the Future of the Non-Proliferation Treaty: A Time for Reassessment, held at York University, Toronto, in May 1985. The papers discuss such topics as the legitimacy, capability, and future of the treaty; the nonproliferation policies of non-nuclear-weapon states; the nuclear industry and the treaty; various aspects of nuclear nonproliferation policies and perspectives; and a review of the Treaty on the Non-Proliferation of Nuclear Weapons. A list of contributors and participants in the conference is given.

Goldblat, Josef, ed. **Non-Proliferation: The Why and the Wherefore.** Stockholm International Peace Research Institute (SIPRI). London: Taylor and Francis Ltd., 1985. 343p. ISBN 0-85066-304-0.

Examines why non-nuclear-weapon states that participate in nuclear activities favor the acquisition of nuclear weapons.

Jones, Rodney W., and others. **The Nuclear Suppliers and Nonproliferation: International Policy Choices.** Lexington, MA: Lexington Books, 1985. 253p. ISBN 0-669-10097-8.

Compilation of papers from a seminar sponsored by the Georgetown University Center for Strategic and International Studies and other organizations, held June 28–29, 1984, in Washington, D.C. The first part of the book deals with theoretical perspectives on nonproliferation. The second contains perspectives from people from nuclear supplier countries, giving their national policies. The third dwells on the problems of nonproliferation for nuclear suppliers. The fourth takes up industry and government cooperation on nuclear supply issues. The fifth part focuses on countries that oppose the principles of nonproliferation and the Non-Proliferation Treaty (NPT). The last part looks at planning for the NPT review conference, the issues that might be raised, and requirements for the future of nonproliferation.

Paranjpe, Shrikant. **U.S. Nonproliferation Policy in Action: South Asia.** New Delhi, India: Sterling Publishers Private Ltd., 1987. 142p. ISBN 81-207-0725-7.

Deals with U.S. nonproliferation policy with special reference to India and Pakistan.

Pilat, Joseph F., ed. **The Nonproliferation Predicament.** New Brunswick, NJ: Transaction Books, Rutgers—The State University, 1985. 135p. ISBN 0-88738-047-6.

Examines the current state of U.S. nonproliferation policy and its effectiveness in the 1980s. Two schools of thought relating to the proliferation of nuclear supply are looked at. On the one hand there are those who feel that nuclear weapons proliferation is directly linked to the spread of nuclear material, technology, and equipment and should be restricted. On the other hand there are those who feel that proliferation is a political problem and seek to improve the supply and security assurances and establish new institutional means to do so. In other words, some feel the current policies are adequate while others feel there should be new initiatives in the nuclear energy or arms control areas.

Reiss, Mitchell. **Without the Bomb: The Politics of Nuclear Nonproliferation.** New York: Columbia University Press, 1988. 337p. ISBN 0-231-06438-1.

Gives a historical view of the international nonproliferation movement since World War II and how perspectives have changed. The author selected six countries—Sweden, South Korea, Japan, Israel, South Africa, and India—that had the ability to develop their nuclear weapons but decided against it and explained their reasons. Special attention is given to political motivations for nuclear weapons acquisition.

Subramanian, Ram R. **Nuclear Competition in South Asia and U.S. Policy.** Berkeley, CA: Institute of International Studies, University of California, 1987. 62p. ISBN 0-87725-530-x.

Discusses American foreign policy concerning nuclear nonproliferation in South Asia. Looks at the consequences of nuclear arms competition between India and Pakistan.

Nuclear Proliferation

Bhatia, Shyam. **Nuclear Rivals in the Middle East.** London: Routledge and Kegan Paul Ltd., 1988. 119p. ISBN 0-415-00479-9.

Gives the evolution of nuclear research and development. Discusses ways that proliferation can be slowed down.

Brito, Dagobert L., Michael D. Intriligator, and Adele E. Wick, eds. **Strategies for Managing Nuclear Proliferation: Economic and Political Issues.** Lexington, MA: Lexington Books, 1983. 311p. ISBN 0-669-06442-4.

Proceedings of a conference held at Tulane University, April 23–25, 1982. Prepared under the auspices of the Center for International and Strategic Affairs, University of California, Los Angeles. Some of the topics examined in this volume are: economics of nuclear power and proliferation; economic development and proliferation; proliferation and stability; proliferation in the Middle East and South Asia; and forecasting and managing proliferation. Biographical sketches of the contributors and editors are included.

Dunn, Lewis A. **Controlling the Bomb: Nuclear Proliferation in the 1980's.** A Twentieth Century Fund Report. New Haven, CT: Yale University Press, 1982. 209p. ISBN 0-300-02820-2.

American efforts to contain proliferation have in recent years focused on preventing countries from acquiring the capability to make nuclear weapons. But such efforts no longer prove adequate, because technology and resources are freely traded. Dunn examines ways in which proliferation can be kept under control by limiting the size and the sophistication of nuclear forces. Steps should be taken to reduce incentives to "go nuclear" by enhancing the security of countries and minimizing uncertainties among rival countries as to each other's intentions about nuclear weapons.

Markey, Edward J., and Douglas C. Waller. **Nuclear Peril: The Politics of Proliferation.** Cambridge, MA: Ballinger Publishing Co., 1982. 183p. ISBN 0-88410-892-9.

The authors stress the danger of the atomic age and the spread of nuclear weapons. Prevention of nuclear war is the greatest issue of all time. Markey, a congressman from Massachusetts, is a leader in the fight to control the atom. He examines the causes of nuclear proliferation and the relationship to the arms race. Part one discusses nuclear safety and atoms for peace; part two discusses nuclear politics Carter style; and part three discusses nuclear politics Reagan style. The author concludes by giving four steps to implement denuclearization: (1) the United States should be the nonproliferation leader; (2) the battle against horizontal proliferation must proceed hand in hand with halting vertical proliferation; (3) the United States should get out of the nuclear export business; and (4) the United States should begin the dismantlement of nuclear power at home.

Snyder, Jed C., and Samuel F. Wells, Jr., eds. **Limiting Nuclear Proliferation.** Cambridge, MA: Ballinger Publishing Co., 1985. 361p. ISBN 0-88730-042-1.

Focuses on the threshold states—Iraq, Israel, India, Pakistan, Argentina, Brazil, and South Africa—and how the spread of nuclear weapons can be limited, stressing the importance of the problem of nuclear proliferation. United States must join other nations to help restrict the spread of nuclear weapons.

Nuclear Energy

General Works

Babin, Ronald. **The Nuclear Power Game.** Translated by Ted Richmond. Montreal: Black Rose Books, 1985. 236p. ISBN 0-920057-30-6.

This book gives the history of the internal dynamics of the Canadian atomic power industry. The second part analyzes the nature of the opposition to the industry. The antinuclear movement developed in two stages. The first stage, lasting from the beginning of the 1970s to the 1979 accident at Three Mile Island, was one of rapid growth. In the second stage, dating from the Three Mile Island accident, the movement adopted a more flexible attitude and tried to become a political force by seeking alliances with other progressive social movements.

Freeman, Leslie J. **Nuclear Witnesses: Insiders Speak Out.** New York: Norton, 1981. 330p. ISBN 0-393-01456-8.

In this book, physicists, carpenters, pipefitters, lab technicians, medical researchers, and others who had direct contact with the nuclear establishment discuss what they learned about nuclear energy. The author interviewed each person to get his reaction to his working conditions. Most of the interviewees were discouraged with the unsafe conditions under which they worked. They had to fight the myth that nuclear power was "safe" and existed for the benefit of the people and that nuclear weapons were necessary for national security. They wanted to express their opinions in order to help other people.

Freudenburg, William R., and Eugene A. Rosa, eds. **Public Reactions to Nuclear Power: Are There Critical Masses?** AAAS Selected Symposium 93. Boulder, CO: Westview Press, 1984. 370p. ISBN 0-86531-708-9.

Based on a symposium held at the January 3–8, 1982, AAAS annual meeting in Washington, DC, sponsored by the Rural Sociological Society, the American Sociological Association and AAAS. Book deals with the marches and demonstrations opposing the nuclear power industry. It is written in a readable style giving public attitudes toward nuclear power.

Lovins, A. B., L. H. Lovins, and Leonard Ross. **Energy/War: Breaking the Nuclear Link.** New York: Harper & Row, 1981. 164p. ISBN 0-06-090852-1.

This book has eleven chapters dealing with such topics as nuclear power and nuclear bombs, disguises and safeguards, collapse of nuclear power, how to save oil, the fatuity of U.S. policy, implementing nonnuclear futures, the nuclear arms race, nuclear power and developing countries, and in the final chapter a once-in-a-lifetime opportunity. Nuclear power is the slowest, most costly, and most dangerous way to displace oil. A condensation of this book appeared in *Foreign Affairs,* Summer 1980.

Mazuzan, George T., and J. Samuel Walker. **Controlling the Atom: The Beginnings of Nuclear Regulation, 1946–62.** Berkeley, CA: University of California Press, 1985. 530p. ISBN 0-520-05182-3.

Traces the early history of nuclear power regulations in the United States when the Atomic Energy Commission was responsible for the protection and safety of people in the use of nuclear energy. The Nuclear Regulatory Commission (NRC) now assumes this responsibility. The book gives the NRC and the public information on the historical antecedents and background of regulatory issues. An appendix gives the chronology of regulatory history, covering the period 1945 through 1963.

Murray, Raymond L. **Nuclear Energy.** 3rd ed. Pergamon Unified Engineering Series, vol. 22. New York: Pergamon Press, 1988. 350p. ISBN 0-08-031628-X.

Provides a factual description of basic nuclear phenomena. Not only are the devices and processes described, but attention is also given to the problems and opportunities that are inherent in a nuclear age. The material is designed for those who want to know about the role of nuclear energy in our society. The basic concepts are presented in a "real world" context. The sequences of presentation proceed from fundamental facts and principles through a variety of nuclear devices to the relation between nuclear energy and peaceful applications.

Platt, A. M., J. V. Robinson, and O. F. Hill, eds. **Nuclear Fact Book.** 2d ed. London: Harwood Academic Publishers, 1985. 192p. ISBN 3-7186-0273-3.

The *Nuclear Fact Book* provides concise summaries and essential facts in the areas of energy production, costs, consumption, nuclear energy production, fuel cycle, and wastes. This is a guide to waste management costs and identifies high-level radioactive wastes and regulations on transportation of nuclear materials.

Scheinman, Lawrence. **The International Atomic Energy Agency and World Nuclear Order.** Washington, DC: Resources for the Future, 1987. 320p. ISBN 0-915707-35-7.

Gives the structure and function of the International Atomic Energy Agency and presents some of the main issues facing the agency.

Shrader-Frechette, K. S. **Nuclear Power and Public Policy: The Social and Ethical Problems of Fission Technology.** Boston, MA: D. Reidel, 1980. 176p. ISBN 90-277-1054-6.

This book grew out of projects funded by the Kentucky Humanities Council and the Environmental Protection Agency. It gives the history of using atomic fission to generate electricity and describes the principal features of a nuclear reactor. Author assumes that reactors will operate normally but questions how the radioactive waste can be stored without causing problems. A core melt catastrophe is discussed showing why public policy regarding liability coverage violates constitutional principles. Author evaluates the cost-benefit of nuclear technology and claims that economically it is both illogical and unethical. The last chapter deals with nuclear safety and the naturalistic fallacy. Policy-making procedures that should be followed in dealing with nuclear technology are outlined.

Smart, Ian, ed. **World Nuclear Energy: Toward a Bargain of Confidence.** Baltimore, MD: Johns Hopkins University Press, 1982. 394p. ISBN 0-8018-2652-7.

Prepared under the auspices of the Royal Institute of International Affairs and based on material published by the International Consultative Group on Nuclear Energy (ICGNE). ICGNE no longer exists, but this book is a composite of some work done by the group. Such topics as peaceful nuclear relations, nuclear energy and international cooperation, nuclear nonproliferation, international safeguards, viability of the civil nuclear industry, and international custody of plutonium stocks are discussed. Many figures and tables are presented. A lengthy glossary and abbreviations are included.

Union of Concerned Scientists. **The Nuclear Fuel Cycle.** MIT Press Environmental Studies Series. Cambridge, MA: MIT Press, 1975. 291p. ISBN 0-262-21005-3.

A worldwide controversy has developed over the safety, environmental, and national security impact of the expanding program of utilizing nuclear reactors to generate electricity. This volume presents fundamental information on the development of nuclear power in an essentially nontechnical form. Topics discussed include radiation-induced lung cancer among uranium miners, radiation hazards associated with uranium mill operations, catastrophic nuclear reactor accidents, nuclear safeguard problems, transportation of radioactive wastes, nuclear fuel reprocessing, and storage and disposal of high-level radioactive wastes.

Weart, Spencer R. **Nuclear Fear: A History of Images.** Cambridge, MA: Harvard University Press, 1988. 535p. ISBN 0-674-62835-7.

Weart, a physicist-historian, describes how Americans have responded to the world of nuclear energy and its militarization. A central theme of the book is the analogy between alchemy and the transmutation of elements on which nuclear energy is founded. The book gives the history of nuclear energy and nuclear warfare with psychological aspects. The antinuclear movement is treated.

Accidents

The Accident at Three Mile Island: The Need for Change: The Legacy of TMI. New York: Pergamon Press, 1979. 199p. ISBN 0-08-025946-4.

This is a report of the 12-member presidential commission that was charged with studying and investigating the accident involving the nuclear power facility on Three Mile Island in Pennsylvania. Book has many pictures of the facility, evacuation zones, and a useful glossary. Biographies of the 12-member commission are an added feature.

Cantelon, Philip L., and Robert C. Williams. **Crisis Contained: The Department of Energy at Three Mile Island.** Social Science and International Affairs Series. Carbondale, IL: Southern Illinois University Press, 1982. 213p. ISBN 0-8093-1079-1.

The Department of Energy's activities at Three Mile Island are presented on a daily and sometimes hourly basis. The book shows how DOE responded quickly to the accident and how the many public agencies at Three Mile Island didn't know what the others were doing or what their responsibilities were. With this accident the public trust in the government and the nuclear industry failed, thus changing the public's idea of the future of the nuclear industry in the United States. The story of Three Mile Island is one of human and institutional behavior as well as nuclear technology.

Ford, Daniel F. **Three Mile Island: Thirty Minutes to Meltdown.** New York: Viking Press, 1982. 271p. ISBN 0-670-70859-3.

Ford gained background for this book by working for several years with the Union of Concerned Scientists. Book tells the story of the Three Mile Island accident from the beginning to the efforts to clean up the plant. Ford discusses the fact that no qualified nuclear engineer, college graduate, nor anyone with technical training who knew what to do in such an emergency was on duty. When the red light came on, the operators on duty knew they were in trouble. State police, Pennsylvania Bureau of Radiological Health, U.S. Department of Energy, Dauphin County Civil Defense Agency, and other emergency agencies were notified of the emergency and requested help. The Nuclear Regulatory Commission office outside Philadelphia could not be reached by telephone, so a message was left with the answering service. NRC officials reacted casually at first to the message that they had a real emergency at Three Mile Island. Manufacturers were sought who produced potassium iodide, approved by the FDA for use in radiation emergencies. Cleanup efforts have not been successful.

Gale, Robert Peter. **Final Warning: The Legacy of Chernobyl.** New York: Warner Books, 1988. 230p. ISBN 0-446-51409-8.

Dr. Gale, the American doctor who led an international radiation treatment team for victims of the Chernobyl accident, discusses what we have learned since that disaster. He gives a firsthand account of how he tried to save Soviet lives and his meeting with Gorbachev.

Martin, Daniel. **Three Mile Island: Prologue or Epilogue?** Cambridge, MA: Ballinger Publishing Company, 1980. 251p. ISBN 0-88410-629-6.

This book about Three Mile Island discusses a reactor in search of a home; structure of the TMI reactor; a transient gone awry; NRC response and what actions were taken. Harold Denton, member of NRC's Emergency Management Team, went to the plant and was the official spokesman on the condition of the reactor. Book is written for the layman. Interesting features of the book are the chronology of the crisis, the technical glossary, and acronyms used in the book. Diagrams of the reactor are helpful.

Stephens, Mark. **Three Mile Island.** New York: Random House, 1980. 245p. ISBN 0-394-51092-5.

The purpose of the book is to tell what really happened at Three Mile Island. It shows what was happening in the reactor, how utility and public officials responded, and the part played by the press. This accident exposed people to excessive doses of radiation because the people in charge put politics, pride, and economics before safety and public health. The author thinks that many of the mistakes that caused the accident could have been prevented. The book contains an extensive glossary.

Environmental and Health Aspects

Eichholz, Geoffrey G. **Environmental Aspects of Nuclear Power.** Ann Arbor, MI: Ann Arbor Science Publishers, 1976. 683p. ISBN 0-250-40138-X.

The book collects essential technical facts about the environmental effects of the nuclear industry to help train and educate scientists and engineers. This is an outgrowth of lecture courses and discussion groups at Georgia Tech. Such topics as nuclear power plants, radiation effects, treatment of radioactive effluents, siting of power plants, transportation of nuclear materials, and radioactive waste disposal are discussed. The Appendix provides a list of elements, symbols, abbreviations, and conversion tables. The book is illustrated with 123 figures and 183 tables.

El-Hinnawi, E. E., ed. **Nuclear Energy and the Environment.** Environmental Sciences and Applications Series, vol. 11. Elmsford, NY: Pergamon Press, 1980. 300p. ISBN 0-08-024472-6.

This book is an outgrowth of a study made by the United Nations Environment Programme concerning the environmental impacts of nuclear energy on production, transport, and use of different sources of energy. The contributors to the book are mainly from Canada, Australia, the United Kingdom, and France. The editor is from Kenya. There are excellent chapters on reactor accidents, nuclear waste disposal, environmental impact of mining and milling of radioactive ores, and transportation of radioactive materials. The editor wrote the chapter on the future of nuclear energy.

Glasstone, Samuel, and Walter H. Jordan. **Nuclear Power and Its Environmental Effects.** La Grange Park, IL: American Nuclear Society, 1980. 395p. ISBN 0-89448-022-7.

This book was written to serve as a summary of safety measures and environmental concerns for the people. Environmental effects associated with nuclear power operations from the mining of uranium to the disposal of waste products are discussed. Safety measures to take when radioactive substances are released into the environment and the potential biological effects of radiation are examined.

International Atomic Energy Agency. **Environmental Aspects of Nuclear Power Stations.** Proceedings of a Symposium Held by the International Atomic Energy Agency in Co-operation with the United States Atomic Energy Commission in New York, 10–14 August 1970. Proceedings Series. Vienna, Austria: 1971. 970p. No ISBN.

This is a collection of papers given at the symposium on environmental aspects of nuclear power stations in New York. It was attended by 350 participants from 25 countries and 9 international organizations. Such topics as siting of power stations, needs for electrical power, standards for control of effluents, and the part played by the public in power plant siting are examined. A comparison is made of the effects of fuel transportation, waste heat discharge, smoke and nonnuclear pollutant emissions, aesthetic requirements, and radioactive emissions in connection with nuclear and fossil-fuel power stations. The book ends with a section on "Prospects for the Future." A list of participants and author index are included.

National Research Council. Panel on the Implementation Requirements of Environmental Radiation Standards. Committee on Radioactive Waste Management. **Implementation of Long-Term Environmental Radiation Standards: The Issue of Verification: A Report.** Washington, DC: National Academy of Sciences, 1979. 65p. ISBN 0-309-02879-5.

This panel conducted a study to find a procedure that would provide assurance that the long-term capacity of a radioactive waste repository would comply with EPA standards. Its final report addressed requirements for assuring that environmental radiation standards have been implemented. The panel concluded that the emplacement of radioactive wastes in properly designed repositories located below the surface in selected geologic formations would reduce the release of radionuclides to a level that would meet the environmental radiation standards. The panel recommended that (1) a procedure be adopted to ascertain whether a geologic site and the construction of a repository would provide adequate isolation of radioactive wastes to satisfy EPA standards; (2) this procedure be a continuous process that evaluates site suitability and satisfactory repository performance before construction begins; and (3) a study be conducted of the capability of the geologic formation to contain the wastes.

Pennsylvania Governor's Fact Finding Committee. **Shippingport Nuclear Power Station: Alleged Health Effects.** Harrisburg, PA: 1974. 132p. No ISBN.

The committee investigated and made recommendations regarding the allegations that releases of radioactive material from the Shippingport Atomic Power Station caused increases in certain diseases and mortality in the surrounding area. The committee recommended a better monitoring program be established; adequate health physics programs supervised by a certified health physicist be initiated by all nuclear power plants; epidemiological laboratory studies be conducted to assess the effects of low-level radiation on human health; better methods for releasing mortality data be developed; and a statewide system for proper evaluation and management of cancer related health problems be implemented. The book has many tables and diagrams and a bibliography.

Pollock, Cynthia. **Decommissioning: Nuclear Power's Missing Link.** Worldwatch Paper 69. Washington, DC: Worldwatch Institute, 1986. 54p. ISBN 0-916468-70-4.

Decommissioning—cleaning up and burying a nuclear plant no longer in use to protect the public from radioactivity—is an essential step in the use of nuclear power. The decommissioning procedure is both costly and complex. Nuclear plants cannot simply be abandoned at the end of their lives or just demolished with a wrecking ball. The biosphere must be kept free of the toxicity of high-level wastes. Commercial reactors should build a decommissioning fund to be used when the plant is shut down so the burden of cost of decommissioning is not borne by the electricity customers and taxpayers. Existing plants must eventually be scrapped.

Radioactive Waste Management and Disposal

Alford, Paula N., and Andrea N. Dravo. **Hot Stuff: Issues in the Management of High-Level Radioactive Waste.** Washington, DC: National League of Cities, 1986. 67p. ISBN 0-933729-15-4.

Addresses problems faced by local governments in the siting, disposal, and transportation of high-level radioactive waste. It discusses the background of the Nuclear Waste Policy Act with its policy issues and the Price-Anderson Act of 1957, which was to protect the nuclear industry from catastrophic risks and activities involving high-level wastes.

Barlett, Donald L., and James B. Steele. **Forevermore: Nuclear Waste in America.** New York: W. W. Norton & Company, 1985. 352p. ISBN 0-393-01920-9.

Forevermore is the result of a series of articles that appeared in the *Philadelphia Inquirer,* November 1983. The authors traveled more than 20,000 miles interviewing people and collecting 125,000 pages of documents on the subject of waste. They found many statistics on nuclear waste but also many irregularities as kept by the National Regulatory Commission. They cite the used fuel assemblies from nuclear reactors, which account for most of the waste in terms of radioactivity, where the statistics expanded or contracted at the statistician's will. Even if no new nuclear plant is opened, the curies will continue to accumulate and the prediction is that by the year 2000 waste stockpiles will be more than 42 million curies—enough to kill everyone on the face of the earth. The book is illustrated with pictures and tables. Supplementary notes and a bibliography are most useful.

Birkhoff, G., and others. **Management of Plutonium Contaminated Waste.** Edited by J. R. Grover. Radioactive Waste Management, vol. 3. New York: Harwood Academic Publishers for the Commission of the European Communities, 1981. 186p. ISBN 3-7186-0110-9.

This study investigates the plutonium-contaminated wastes coming from a mixed-oxide fabrication plant. These wastes occur as a broad spectrum of materials with quantities and characteristics dependent upon the plant capacity, process routes employed, and the operation philosophy. The current practices for treating the solid and liquid wastes are described, together with monitoring methods, processes for the recovery of plutonium, and current practices of storage and disposal. Special treatment is given to waste minimization. New processes are described. The volume concludes with future management strategies.

Brookins, Douglas G. **Geochemical Aspects of Radioactive Waste Disposal.** New York: Springer-Verlag, 1984. 347p. ISBN 0-387-90916-8.

Brookins attempts to summarize mainly government-sponsored research reports on radioactive waste and its disposal. Background material on radioactive wastes and geotoxicity, geology, and geochemistry are covered for both the geologist and the nuclear scientist. Book focuses mainly on the geochemistry aspect, showing that rocks can probably retain radioactive waste materials and not pose a threat to public health.

Burns, Michael E., ed. **Low-Level Radioactive Waste Regulation: Science, Politics and Fear.** Chelsea, MI: Lewis Publishers, 1988. 311p. ISBN 0-87371-026-6.

Book is the result of three symposia sponsored by the Division of Chemistry and the Law of the American Chemical Society.

Carter, Luther J. **Nuclear Imperatives and Public Trust: Dealing with Radioactive Waste.** Washington, DC: Resources for the Future, 1987. 473p. ISBN 0-915707-29-2.

Discusses the sources of public unease, such as Three Mile Island cleanup and wastes, containment wastes and the fuel cycle, technology involved, and the reprocessing dilemma at such plants as Hanford, Savannah River, and West Valley. The second part deals with the search for a waste policy. The third part examines the international waste problem in Europe and Japan. The last section deals with the action that must be taken to solve the radioactive waste problem.

Chapman, Neil A., and Ian G. McKinley. **The Geological Disposal of Nuclear Waste.** New York: John Wiley & Sons, 1987. 280p. ISBN 0-471-91249-2.

This book is devoted to the idea that once radioactive waste is disposed of it should never return to plague our environment. Since the early 1970s, wastes, especially radioactive, have come under strict regulatory control. The book was written for people with a nonscientific background. Specific references are provided for those who wish additional information. It gives an overall view of geological disposal of radioactive wastes.

Clark, D. E., W. B. White, and A. J. Machiels, eds. **Nuclear Waste Management II.** Proceedings of the Third International Symposium on Ceramics in Nuclear Waste Management Held at the 88th Annual Meeting of the American Ceramic Society, April 28–30, 1986, Chicago, Illinois. *Advances in Ceramics,* vol. 20. Westville, OH: American Ceramic Society, 1986. 773p. ISBN 0-916094-82-0.

Papers from this symposium cover general areas of nuclear waste management; waste generation, management and regulation; waste-form development and engineering scale-up; repository programs; waste form and natural analog chemical durability; and systems testing, including correlation of laboratory and field results. Extensive studies, modeling efforts, and field tests are underway to understand waste forms performance in potential geologic formations. Borosilicate glass is used as the waste form by several countries. The United States will use it to encapsulate high-level wastes stored at the Savannah River Plant in Aiken, South Carolina. Studies are being made to determine the best encapsulant for all wastes. Amorphous and crystalline ceramics are thought to provide a safe and effective barrier against the disposal of radioactive contaminants into the biosphere.

Colglazier, E. William, Jr., ed. **The Politics of Nuclear Waste.** Pergamon Policy Studies on Energy. New York: Pergamon Press, 1982. 264p. ISBN 0-08-026323-2.

Radioactive wastes present a difficult technological problem. Hazardous wastes from plutonium production for military purposes and from electricity generation by commercial reactors will have to be safely managed for centuries. All areas of radioactive waste management—from storage and transport to permanent disposal, from reprocessed waste and spent fuel to low-level waste—abound in social and political and institutional issues. The complexity of these issues is illustrated in the attempts to site a repository. In the United States the federal government is responsible for the permanent disposal of high-level radioactive waste. States and local communities have been reluctant to accept either technical or institutional assurances from the federal government on radioactive waste management.

D'Alessandro, Marco, and Arnold Bonne. **Radioactive Waste Disposal into a Plastic Clay Formation.** Radioactive Waste Management. Chur, Switzerland: Harwood Academic Publishers, 1981. 148p. ISBN 0275-7273.

Since 1974 the Research Center SCK/CEN, Belgium, has studied the suitability of boom clay at Mol as a potential disposal formation for solidified radioactive waste. Research has been performed on site reconnaissance, evaluation of suitability of clay as host rock for radioactive waste, and conceptual repository design. This report identifies some aspects of the geologic formation, presents a range of failure probabilities, and analyzes the relative importance of the most relevent descriptive events able to breach the geological containment. This is a progress report reflecting the state of the present knowledge. More statistical data are required for a definitive analysis.

Deese, David A. **Nuclear Power and Radioactive Waste: A Sub-Seabed Disposal Option?** Lexington, MA: Lexington Books, 1978. 206p. ISBN 0-669-02114-8.

Discusses such problems as past marine disposal practice—the freedom to pollute; national regulatory postures and effects on sub-seabed disposal of radioactive waste; international law regarding sub-seabed disposal of radioactive waste; political pressures; ethical considerations; and institutional possibilities for management and prevention of sub-seabed disposal of radioactive wastes. The book contains many figures and tables. A list of abbreviations and a selected bibliography are useful.

Dlouhý, Zdeněk. **Disposal of Radioactive Wastes.** Studies in Environmental Science, 15. Translated from the Czech. New York: Elsevier Scientific, 1982. 264p. ISBN 0-444-41696-X.

The growing demands on electric power and lack of power resources have made nuclear power important. Nuclear power production has become one of the fastest-growing industries, thus the problem of safe disposal of radioactive wastes that accompany the whole fuel cycle from uranium mining and fuel processing to power production and the reprocessing of spent fuel have been brought to the fore because of stringent environmental laws.

Eid, C., and A. J. Van Loon, eds. **Incineration of Radioactive Waste.** London: Graham & Trotman, 1985. 205p. ISBN 0-86010-677-2.

The nuclear fuel cycle, together with the use of radioisotopes in industry and medical treatment, generates a variety of low- and intermediate level solid radioactive wastes. A large proportion of this waste is combustible and can be reduced substantially in volume and weight by incineration. In addition, incineration treatment converts the waste to a form well suited for management. This volume presents the findings of a seminar organized by the Commission of the European Communities in 1984. The major sections include incineration of solid and other radioactive wastes, measurement and control, and off-gas filtration and release. The material discusses modern technology and the future improvement of safety through output capacity and reduction of secondary waste.

Frye, Alton. **The Hazards of Atomic Wastes: Perspectives and Proposals on Ocean Disposal.** Washington, DC: Public Affairs Press, 1962. 45p. No ISBN.

The purpose of this book is to examine the problem of disposal of radioactive wastes. Measures must be taken to control the dangers of polluting the oceans to prevent serious effects on marine resources and restricting the use of oceans for many activities. Radioactive waste looms as a serious problem in the atomic age. This is a small book but does discuss radiation dangers, the ocean as a depository of wastes, and international law and oceanic pollution and makes some proposals to help solve the problems.

Glueckauf, E., ed. **Atomic Energy Waste: Its Nature, Use and Disposal.** New York: Interscience Publishers, 1961. 420p. No ISBN.

This book discusses the problems faced in the early years of atomic energy. Tremendous developments have been made in the treatment and storage of waste products over the years.

International Atomic Energy Agency. **Acceptance Criteria for Disposal of Radioactive Wastes in Shallow Ground and Rock Cavities.** Safety Series, Recommendations. Vienna, Austria, 1985. 38p. ISBN 92-00-123985-8.

This book addresses underground disposal—in shallow ground, rock cavities, or deep geological formations—as a safe disposition of radioactive waste so man and other living species will not be detrimentally affected at the present or in the future. The effective and safe isolation of radioactive waste depends on the overall performance of the site, the repository, and the waste package.

______. **Concepts and Examples of Safety Analyses for Radioactive Waste Repositories in Continental Geological Formations.** Safety Series, no. 58. Vienna, Austria, 1983. 171p. ISBN 92-0-123383-3.

This document is for use by specialists responsible for planning, performing, and reviewing assessments of underground radioactive waste repositories. It summarizes safety analyses. Guidelines include such things as selection of repository sites, waste acceptance criteria, design and construction of repositories, safety steps, and operation, surveillance, and shutdown of repositories.

______. **Control of Radioactive Waste Disposal into the Marine Environment.** Safety Series, no. 61. Vienna, Austria, 1983. 128p. ISBN 92-0-123883-5.

This volume gives information on the background, technical procedures, and control of marine disposal of radioactive wastes and on international law and international recommendations. The first international meeting on radioactive waste disposal into the ocean was held in Geneva in 1958 at the U.N. Conference on the Law of the Sea, where an agreement was made to prevent pollution of the seas by the dumping of radioactive waste. Suitable packaging must be used to minimize the release of radioactivity to prevent contamination of the marine environment.

______. **Disposal of Low- and Intermediate-Level Solid Radioactive Wastes in Rock Cavities: A Guidebook.** Safety Series, no. 59. Vienna, Austria, 1983. 49p. ISBN 92-0-123483-X.

Provides guidelines for people practicing disposal of low- and intermediate-level radioactive waste in rock cavities. They take into account local conditions, such as natural circumstances, characteristics of the waste, and national and international regulations. An appendix gives a synopsis of national activities on rock cavity disposal concepts.

______. **Disposal of Radioactive Wastes into Rivers, Lakes and Estuaries.** Report of a Panel Sponsored by the IAEA and the World Health Organization. Safety Series, no. 36. Vienna, Austria, 1971. 77p. No ISBN.

The book discusses the use of rivers not only as a source of freshwater supplies but also as repositories for effluents from industry and sewage treatment works—water that is used over and over again. To use water resources efficiently, their quantity must be conserved and pollution must be prevented. The mechanisms and parameters that affect and control the fate of radionuclides introduced into fresh waters are discussed.

______. **Management of Radioactive Wastes from Nuclear Power Plants: Code of Practice.** Safety Series, no. 69. Vienna, Austria, 1985. 34p. ISBN 92-0-123685-9.

The Code of Practice defines the requirements for the management of radioactive wastes from thermal neutron power plants and emphasizes safety requirements for safe operation of nuclear power plants, radiological safety, environmental protection, and waste management practices.

______. **Management of Wastes from the Mining and Milling of Uranium and Thorium Ores: A Code of Practice and Guide to the Code.** Safety Series, no. 44. Vienna, Austria, 1976. 50p. ISBN 92-0-123276-4.

The first part of the book is the Code of Practice approved by the Board of Governors of the IAEA as part of the agency's safety standards. The second part of the book, the Guide to the Code, shows how requirements of the code can be met.

______. **The Oceanographic and Radiological Basis for the Definition of High-Level Wastes Unsuitable for Dumping at Sea.** Safety Series, no. 66. Vienna, Austria, 1984. 50p. ISBN 92-0-623384-X.

This document is concerned with the selection of parameters for calculating dumping rate limits so there are no ill effects on the marine ecosystems. It also mentions a category of material that may not be dumped but does not spell out materials that can be dumped. A list of references to be considered is given for added information. Evaluates the impact of mining of manganese from the sea bottom and the use of plankton and deep-sea fish as food in the human diet.

______. **Performance Assessment for Underground Radioactive Waste Disposal Systems.** Safety Series, no. 68. Vienna, Austria, 1985. 37p. ISBN 92-0-123485-6.

This book provides valuable information for specialists responsible for making and reviewing performance assessments of underground disposal systems for radioactive wastes. It provides insights into the methods employed in performance assessment work and gives practical guidance in specific technical areas of importance. The book focuses on emplacement of solid wastes in deep geological repositories, in man-made repositories, in rock cavities, and in shallow-ground repositories.

______. **Radioactive Waste Disposal into the Sea.** Report of the Ad Hoc Panel Convened by the Director of the IAEA. Safety Series, no. 5. Vienna, Austria, 1961. 174p. No ISBN.

This report deals with contamination of the sea from disposal of radioactive waste. It was written for people who are interested in or have responsibility for the problem. Appendices are found in the book for the technically concerned person to use in dealing with waste disposal operations.

______. **Safety Analysis Methodologies for Radioactive Waste Repositories in Shallow Ground.** Safety Series, no. 64. Vienna, Austria, 1984. 53p. ISBN 92-0-123484-8.

Underground disposal of wastes, appropriately immobilized or packaged, seems to be an adequate way of providing protection for humans and the environment. Four types of underground disposal repository systems are in use or being developed. Three involve emplacement of solid wastes in shallow-ground repositories, in man-made or natural rock cavity repositories, and in deep geological repositories. The other involves injection of self-solidifying fuels containing wastes into fractures within impermeable strata. This report deals only with shallow-ground repositories.

______. **Safety Assessment for the Underground Disposal of Radioactive Wastes.** Safety Series, no. 56. Vienna, Austria, 1981. 46p. ISBN 92-0-623181-2.

This document discusses approaches and areas to be considered in making safety assessments of underground radioactive waste repositories, with emphasis on repositories for long-lived radioactive wastes in deep geological formations.

______. **Shallow Ground Disposal of Radioactive Wastes: A Guidebook.** Safety Series, no. 53. Vienna, Austria, 1981. 52p. ISBN 92-0-123281-0.

Guidelines are given for those who are planning the use of shallow-ground disposal for selected radioactive wastes. Guidelines include the following subjects: generic and regulatory activities and safety measures; investigation and selection of repository sites; waste acceptance criteria; design and construction of repositories and operation; and shut-down and surveillance of repositories. A list of other books is given for consultation.

______. **Underground Disposal of Radioactive Wastes.** Safety Series, no. 54. Vienna, Austria, 1981. 56p. ISBN 92-0-123381-7.

This volume examines the appropriate system for the management and disposal of radioactive waste that is a result of the use of nuclear power. Underground disposal that immobilizes the waste should protect man and environment from the potential hazard radioactive wastes pose.

International Nuclear Fuel Cycle Evaluation. **Waste Management and Disposal, Report of INFCE Working Group 7.** Vienna, Austria: International Atomic Energy Agency, 1980. 288p. ISBN 92-0-159780-0.

This study examines the legal and institutional aspects of nuclear waste management and disposal. The primary consideration in nuclear waste management and disposal is to protect the public from the radioactivity of waste materials. Also considered are the effects on the environment and costs of waste management and disposal.

Jackson, Thomas C., ed. **Nuclear Waste Management: The Ocean Alternative.** Sponsored by the Oceanic Society in Cooperation with the Georgetown University Law Center Institute for International and Foreign Trade Law and the Center for Law and Social Policy. Pergamon Policy Studies on Energy. New York: Pergamon Press, 1981. 124p. ISBN 0-08-027204-5.

This book deals with the disposal of radioactive waste in the ocean. Some scientists feel that problems posed by radioactive waste are insurmountable and an end should be put to the commercial atomic generating plants, yet others see an increased expansion of such plants. The book focuses on the "ocean alternative" to be certain marine disposal will never be used just as the convenient method of disposing of nuclear waste.

Johansson, Thomas B., and Peter Steen. **Radioactive Waste from Nuclear Power Plants.** Berkeley, CA: University of California Press, 1981. 197p. ISBN 0-520-04199-2.

The Nuclear Stipulation Act was passed by the Swedish Parliament in 1977 and imposed stringent rules regarding the long-term management of highly radioactive wastes from nuclear power plants. A method for the management of wastes had to be presented before permission to load and operate any new reactors could be given. Late in 1977 the Swedish nuclear industry presented a plan for the management of vitrified liquid wastes from reprocessed spent fuel.

Kasperson, Roger E., and others, eds. **Equity Issues in Radioactive Waste Management.** Cambridge, MA: Oelgeschlager, Gunn & Hain, 1983. 381p. ISBN 0-89946-055-0.

Few technologies have caused as intense public concern as nuclear energy. This book is by a group of scholars sponsored by the National Science Foundation who studied the problems of equity-locus, legacy, and labor-laity issues. The intent of the book is to provide analyses that inform current public policy deliberations and planning. The authors make proposals for a more equitable approach to the management of radioactive waste. The book is divided into five parts. Part 1 gives a perspective, Part 2 discusses the locus problem, Part 3 examines the legacy problem, Part 4 discusses the labor problem, and Part 5 deals with equity in radioactive waste management. Each chapter has a list of references.

Krischer, W., and R. Simon, eds. **Testing, Evaluation and Shallow Land Burial of Low and Medium Radioactive Waste Forms.** Proceedings of a Seminar Organized by the European Communities on Radioactive Waste Management and Disposal Held in Geel, Belgium, 28–29 September 1983. Paris: Harwood Academic Publishers, 1984. 231p. ISBN 3-7186-0206-7.

Since the beginning of nuclear energy, radioactive waste management has been a problem. This volume deals with the processing and disposal of such waste form in a safe manner. Also surveyed is the results of an analysis of the present and future risks of disposing of low-activity wastes at shallow depths.

Lau, Foo-Sun. **Radioactivity and Nuclear Waste Disposal.** Research Studies in Nuclear Technology. New York: John Wiley & Sons, 1987. 614p. ISBN 0-471-91524-6.

The management of radioactive wastes has given rise to one of the major technical and political arguments of the present day. Legislative and regulatory bodies of most countries with nuclear power programs have largely taken the position that reactor operations today should not constitute a burden on future generations. The purpose of the book is to provide a background to nuclear waste disposal. The first part of the book reviews nuclear theory and associated radioactivity. The development of radiation for medical therapy and diagnosis is discussed. Other topics discussed in part two include waste management, properties of radioactive materials, classification of radioactive wastes, radioactive wastes from nuclear reactors, disposal options, selection of sites, and waste packaging. A major feature of the volume is the 74 figures and 130 tables.

Lindblom, Ulf, and Paul Gnirk. **Nuclear Waste Disposal—Can We Rely on Bedrock?** New York: Pergamon Press, 1982. 60p. ISBN 0-08-027608-3.

This volume deals with the fact that nuclear power reactors produce useful energy in the form of heat and electricity but also produce waste materials that are dangerous to mankind. Nuclear wastes must be controlled so they are not detrimental. This book examines an underground repository that can be backfilled and sealed. The repository site must be selected carefully to be sure it will not have an earthquake or be disturbed by man for thousands of years.

Lipschutz, Ronnie D. **Radioactive Waste: Politics, Technology, and Risk.** Cambridge, MA: Ballinger Publishing Co., 1980. 247p. ISBN 0-88410-621-7.

This book, a report of the Union of Concerned Scientists, provides information about both the technical and nontechnical issues needed for meaningful public involvement in this major issue. It discusses the nature and hazards of radioactivity and production and disposal of radioactive waste. It examines the federal government program and shows obstacles that must be overcome in order to carry out the program. The book contains a glossary and a lengthy bibliography that is broken down by specific subjects.

Majumdar, Shyamal K., and E. Willard Miller, eds. **Management of Radioactive Materials and Wastes: Issues and Progress.** Easton, PA: The Pennsylvania Academy of Science, 1985. 405p. ISBN 0-9606670-4-0.

The problem of the disposal of radioactive wastes is one that has reached a critical dimension in our time. This volume is divided into five parts. Part one considers types, sources, and management of radioactive wastes and covers such aspects as classification of radioactive wastes, low-level waste management, and shallow land-burial of radioactive materials. The storage, transportation, and disposal problems of radioactive materials are discussed in part two. Part three covers sociopolitical considerations in disposal of radioactive wastes. Part four considers emergency planning and preparation, including radioactive spills and nuclear accidents. The last part is devoted to radiation standards, the environment, and public health. This book provides valuable information for scientists, engineers, and social scientists.

McCarthy, Gregory J., ed. **Scientific Basis for Nuclear Waste Management.** Proceedings of the Symposium on Science Underlying Radioactive Waste Management Held in Boston, MA, 1978. New York: Plenum Press, 1979. 563p. ISBN 0-306-40181-9.

These proceedings cover such topics as radioactive waste management; radiation and geochemical interactions near the repository; interactions, retention, and migration of radionuclides in geomedia; safety assessment; and treatment and isolation of wastes.

Miles, Edward L., Kai N. Lee, and Elaine M. Carlin. **Nuclear Waste Disposal under the Seabed.** Policy Papers in International Affairs, no. 22. Berkeley, CA: Institute of International Studies, University of California, 1985. 112p. ISBN 0-87725-522-9.

Identifies, evaluates, and strengthens understanding of the policy issues inherent in the sub-seabed disposal option. Technical information is used to inform or clarify policy issues, such as the effect of heat on the near-field sediments in the disposal sites, life of the container, risks in loading canisters on ships and transporting them to the disposal sites, and risk to the marine environment. Also discussed are the domestic political problems for the United States and international problems of creating and operating a sub-seabed disposal.

Milnes, A. G. **Geology and Radwaste.** Orlando, FL: Academic Press, 1985. 328p. ISBN 0-12-498070-8.

The first part of this book is general in scope, outlining the nature and magnitude of the radioactive waste disposal problem and types of solutions proposed. The second part is purely geological in nature. It is written in a nontechnical style to be understood by anyone interested in environmental problems. So many aspects of geology are discussed that a broad overview of the earth sciences emerges. The last part of the book discusses difficulties of applying geological knowledge to such questions as risk assessment and site selection. The problem of repository siting is highly political, and earth scientists find themselves as pawns in an economic-political game, to be used or discarded with little regard for the scientific merits of the case. A lengthy list of references concludes the volume.

Moore, John G., ed. **Scientific Basis for Nuclear Waste Management.** Proceedings of the Symposium on the Scientific Basis for Nuclear Waste Management Held in Boston, MA, 1980. New York: Plenum Press, 1981. 632p. ISBN 0-306-40803-1.

Papers in these proceedings encompass various aspects of high-level and non-high-level radioactive waste management ranging from waste form production and repository characterization to performance assessment, including radiation effects and radionuclide migration.

Murdock, Steven H., F. Larry Leistritz, and Rita R. Hamm, eds. **Nuclear Waste: Socioeconomic Dimensions of Long-Term Storage.** Westview Special Studies in Science, Technology, and Public Policy/Society. Boulder, CO: Westview Press, 1983. 343p. ISBN 0-86531-447-0.

The socioeconomic implications of nuclear waste management and repository siting are examined, as well as attempts to find solutions to the problems. Hazardous waste sites will no doubt be selected in the West and South because of the sparse population and rural areas. Local issues must be resolved while considering long-term interests of the community in siting repositories.

National Council on Radiation Protection and Measurements. **Radioactive Waste.** Proceedings of the Twenty-first Annual Meeting of the National Council on Radiation Protection and Measurements Held April 3–4, 1985, at Bethesda, MD. Bethesda, MD, 1986. 289p. ISBN 0-913392-81-2.

The meeting was concerned with the future of radioactive waste disposal, sources, disposal of uranium mill tailings, management of spent fuel, and many other aspects of radioactive waste.

National Research Council. Committee on Oceanography. **Disposal of Low-Level Radioactive Waste into Pacific Coastal Waters.** National Research Council Publication 985. Washington, DC: National Academy of Sciences, 1962. 87p. No ISBN.

This is a report of a working group of the Committee on Oceanography to consider the problems of disposal of low-level radioactive wastes in the Pacific Ocean off the North American coast. Their problem was to decide by what means and in what amounts it is feasible to dispose of radioactive wastes in the ocean without curtailing the present or future use of the resources of the region. Disposal should not be to the extent that humans would have to be excluded because of radiological hazards, marine organisms would be damaged, or sea products would contain unacceptable levels of radioactive waste to be rejected by people.

National Research Council. Panel on Hanford Wastes. Committee on Radioactive Waste Management. **Radioactive Wastes at the Hanford Reservation, A Technical Review.** Washington, DC: National Academy of Sciences, 1978. 269p. ISBN 0-309-02745-4.

This report is the result of a year's study by the panel, which reviewed the scientific and technological aspects of the management of radioactive wastes at the Hanford Reservation in southeastern Washington. The study was requested by the U.S. Energy Research and Development Administration (ERDA) and the Council on Environmental Quality (CEQ) to evaluate the handling of accumulated wastes from plutonium production for military purposes at Hanford. The panel studied the radiation hazards from Hanford wastes, alternatives for long-term disposal, and effects of accidents, sabotage, climatic, or geologic changes. The book contains a glossary, acronyms, abbreviations, and symbols used as well as a lengthy list of references.

National Research Council. Panel on Land Burial. Committee on Radioactive Waste Management. **The Shallow Land Burial of Low-Level Radioactively Contaminated Solid Waste.** Washington, DC: National Academy of Sciences, 1976. 150p. ISBN 0-309-02535-4.

This is a study of the problems associated with shallow land-burial of solid low-level radioactive waste. The panel reviewed and evaluated the scientific and technical aspects of the shallow land-burial of waste contaminated with low levels of radioactive materials and showed that large-scale social problems of waste management existed. The appendices give summaries of the sites visited, burial sites, and hydrogeologic research.

National Research Council. Panel on Savannah River Wastes. Board on Radioactive Waste Management. **Radioactive Waste Management at the Savannah River Plant: A Technical Review.** Washington, DC: National Academy Press, 1981. 68p. ISBN 0-309-03227-X.

This publication is the result of a panel that examined the effectiveness of current practices for managing low-level and high-level waste and the effectiveness of programs to monitor disposal sites. Also examined is the environmental safety of current and planned practices at disposal sites. The panel conducting the study was to consider such things as cost and hazards and suggest alternative plans at the Savannah River Plant. After several trips to the plant to study the problems, the panel made several recommendations. The book contains a helpful glossary and a lengthy list of references.

National Research Council. Panel on Social and Economic Aspects of Radioactive Waste Management. Board on Radioactive Waste Management. **Social and Economic Aspects of Radioactive Waste Disposal, Considerations for Institutional Management.** Washington, DC: National Academy Press, 1984. 175p. ISBN 0-309-03444-2.

This panel was established at the request of the U.S. Department of Energy (DOE) to study the socioeconomic aspects of nuclear waste repository siting and the operation of a generic radioactive waste repository and to make suggestions for incorporating socioeconomic considerations into the repository selection process. The panel expanded its work to study the transportation and temporary storage of wastes. The panel found that site selection is both a political and social issue. The two conditions that must be met if socioeconomic criteria are to be used in selecting repository sites are: (1) Effects likely to result from choosing one or another option must be specified, in regard to an individual repository and to the deployment of an entire waste management system. (2) The social values relevant to the concerns and goals of different social groups, as they bear on the socioeconomic effects, must be specified.

National Research Council. Waste Isolation Systems Panel. **A Study of the Isolation System for Geologic Disposal of Radioactive Wastes.** Washington, DC: National Academy Press, 1983. 345p. ISBN 0-309-03384-5.

The panel reviewed the alternative technologies available for the isolation of radioactive waste in mined geologic repositories and identified appropriate technical criteria to achieve satisfactory overall performance of a geologic repository. It had to take into account the amount of radiation exposure from the hazardous radionuclides that might be transported from the waste repository to the biosphere. The book provides a lengthy glossary, a list of units and nomenclature, and tables giving the characteristics of waste materials, calculated individual doses per unit concentration of radionuclides released into the environment, and ingestion dose factors for radionuclides.

Organization for Economic Cooperation and Development. **Long-Term Management of Radioactive Waste—Legal, Administrative and Financial Aspects.** Washington, DC: Nuclear Energy Agency, Organization for Economic Cooperation and Development, 1984. 133p. ISBN 92-64-12622-8.

This volume provides a survey of regulatory, technical, and societal issues concerning the management of radioactive waste. It studies the sources and characteristics of a variety of radioactive wastes and describes various waste management options, such as disposal under the ocean floor, geologic disposal, and shallow land-burial. It reviews major issues in radioactive waste management.

Park, P. Kilho, and others. **Radioactive Wastes and the Ocean.** A Wiley Interscience Publication. New York: John Wiley & Sons, 1983. 522p. ISBN 0-471-09770-5.

Presents an overall view of the disposal of waste in the ocean and evaluates the impact of the disposal of waste materials on human life, on marine biota, and on proper uses of the ocean. Because the quantity and kind of waste materials have increased enormously, they must be disposed of in ways that do not destroy our environment. One solution is to recycle waste materials to be beneficial rather than destructive. Waste materials have been disposed of in the marine environment, or in rivers and streams, which finally reach the sea, for years. It is known that bodies of water have been degraded, and disposal of wastes into the hydrosphere must be controlled to preserve the oceans. Resources of the sea must be preserved and protected for the future.

Resnikoff, Marvin. **The Next Nuclear Gamble: Transportation and Storage of Nuclear Waste.** New York: Council on Economic Priorities, 1983. 378p. ISBN 0-87871-020-5.

The entire book deals with the transportation and storage of nuclear waste. The risks, costs, and alternatives for handling irradiated nuclear fuel are examined. Retaining nuclear wastes at reactor sites in dry storage casks until a permanent depository is in operation is recommended. Dry storage casks are the most promising on-site storage facility, and the cheapest as well as the safest, if they are to be used for shipment. The book is well illustrated with figures and tables. Each chapter has a list of references.

Roy, Rustum. **Radioactive Waste Disposal: The Waste Package,** vol. 1. New York: Pergamon Press, 1982. 232p. ISBN 0-08-027541-9.

This is volume 1 of a series of monographs on the future technology of radioactive waste disposal. The book is written for the entire science community interested in radioactive waste, the geologist, the metallurgist, and the nuclear physicist. Major developments in the field are reviewed.

Sarshar, M. M. **Reliability of Sub-Seabed Disposal Operations for High Level Waste.** London: Graham & Trotman Ltd., 1986. 176p. ISBN 0-80610-835-X.

This report was prepared for the European Atomic Energy Community Cost-Sharing Research Programme on Radioactive Waste Management and Disposal. Two methods of disposal, by drilled emplacement and by penetrator as developed by Taylor Woodrow and Ove Arup and Partners, have been analyzed in the study. To assess the reliability, two methods were used: (1) failure mode effects and critical analysis (FMECA) and (2) fault tree analysis (FTA). Two faults of the FTA evolved—loss of lives of people involved in the operation and release of radioactivity into the environment. Of the two methods of disposal, the penetrator method is in general a safer operation, ignoring any past emplacement differences. The book has many tables and figures to help explain the methods. Five appendices contain valuable information.

Shapiro, R.C. **Radwaste.** New York: Random House, 1981. 288p. ISBN 0-394-51159-X.

The author examines uranium-mill tailings, the first radioactive wastes produced, which are increasing in quantity each year. He defines high-level and low-level wastes in great detail. There are also chapters dealing with commercial spent fuel decontamination and decommissioning, transuranic wastes, and transportation of radioactive waste. He believes that legislation should emphasize permanent rather than temporary disposal of radioactive wastes. The book is written in a nontechnical style, making it easy reading.

Simon, R., ed. **Radioactive Waste Management and Disposal.** Proceedings of the Second European Community Conference, Luxembourg, April 22–26, 1985. Cambridge, England: Cambridge University Press, 1986. 734p. ISBN 0-521-32580-3.

This volume presents the proceedings of the Second European Community Conference of 1985. The prime objective of the conference was to explore and find effective means for ensuring the safety of man and his environment against the potential hazards arising from radioactive wastes. The main topics included treatment and conditioning technology; testing and evaluation of waste forms and packages; geologic disposal in salt, granite, and clay formations; and migration and performance analysis of geological isolation systems. This volume provides a comprehensive treatment of waste disposal.

Smith, Thomas P. **A Planner's Guide to Low-Level Radioactive Waste Disposal.** Planning Advisory Service Report, no. 369. Chicago, IL: American Planning Association, 1982. 53p. No ISBN.

This book is aimed to help planners become familiar with issues that they must deal with in managing low-level radioactive and hazardous wastes. Emphasis is on site selection standards for land-burial facilities for low-level radioactive wastes, the types of nuclear wastes, the health risks, the transportation of radioactive materials, and state and local government participation in the disposal of radioactive wastes. A list of films on radioactive waste is found in an appendix. Another appendix includes an annotated bibliography including materials on basic readings, policy, health effects, transportation issues, and technical references.

Stewart, Donald C. **Data for Radioactive Waste Management and Nuclear Applications.** A Wiley-Interscience Publication. New York: John Wiley, 1985. 297p. ISBN 0-471-88627-0.

This book brings together in a single volume information on radioactive waste management useful to government officials, those preparing impact statements, and project managers. The information should be valuable for other nuclear applications as well.

Straub, Conrad P. **Public Health Implications of Radioactive Waste Releases.** Geneva, Switzerland: World Health Organization, 1970. 61p. No ISBN.

This book attempts to designate practices and operations that might result in potential radiation exposure to people through the release of radioactive materials into the environment. Radioactive materials will continue to be used for beneficial purposes. Knowing the characteristics of the waste, the paths the radionuclides take through the environment, the radionuclides involved, and the measures to be taken to limit exposure to radiation will help minimize the hazard to public health. Appendix 1 lists the exposure pathways—how radioactive materials reach man. Appendix 2 lists sources, quantities, and composition of wastes from nuclear operations. Appendix 3 discusses the problem of discharge of radioactive materials into rivers and streams.

Surrey, John, ed. **The Urban Transportation of Irradiated Fuel.** London: Macmillan Press, 1984. 336p. ISBN 0-333-36938-6.

This book is concerned about the transportation of spent nuclear fuel through densely populated areas and the safety steps needed. To reroute away from built-up areas would be the obvious thing to do, except it endangers those who live near the alternative route. Guidelines have been established for safe transport of radioactive materials.

U.S. Congress, Office of Technology Assessment. **Managing the Nation's Commercial High-Level Radioactive Waste.** New York: UNIPUB InfoSource International in Cooperation with the Office, 1985. 348p. ISBN 0-89059-057-5.

Report gives findings and conclusions of the Office of Technology Assessment analysis of federal policy for the management of commercial high-level radioactive waste. It is to contribute to the implementation of the Nuclear Waste Policy Act of 1982. A major conclusion is that the act provides sufficient authority for developing and operating a waste management system for disposal in geologic repositories.

University of Texas. Lyndon B. Johnson School of Public Affairs. **High-level Radioactive Waste and Spent Nuclear Fuel Disposal: An Assessment of Impact Evaluations and Decision-making Systems: A Report by the High-Level Nuclear Waste Disposal Policy Research Project.** Policy Research Report no. 84. Austin, 1987. 165p. ISBN 0-89940-688-2.

This publication focuses on site selection under the Nuclear Waste Policy Act of 1982.

Walker, Charles A., Leroy C. Gould, and Edward J. Woodhouse, eds. **Too Hot to Handle? Social and Policy Issues in the Management of Radioactive Wastes.** New Haven, CT: Yale University Press, 1983. 209p. ISBN 0-300-02899-7.

The world's first atomic bomb exploded on July 16, 1945, in a desert in New Mexico and began the nuclear age. Ten years later the United States launched the Atoms for Peace program to harness the power of the atom for constructive purposes. This book discusses whether the atom will become man's obedient servant or become man's doom. Another very important concern is making sure nuclear power plants are safe and developing safe procedures to dispose of radioactive wastes that are produced daily. Nuclear power development in the future will depend upon how radioactive wastes can be managed.

Wicks, George G., and Wayne A. Ross, eds. **Nuclear Waste Management.** Proceedings of the Second International Symposium on Ceramics in Nuclear Waste Management Held at the 85th Annual Meeting of the American

Ceramic Society, April 24–27, 1983, Chicago, Illinois. *Advances in Ceramics*, vol 8. Columbus, OH: American Ceramic Society, 1984. 746p. ISBN 0-916094-55-3.

During this symposium 81 papers were presented by representatives from the United States, Sweden, Germany, Italy, Japan, Canada, Finland, France, Australia, and the United Kingdom. The implication and importance of the Waste Policy Act of 1982 and overviews of defense and commercial waste activities were discussed by representatives from the Department of Energy. Technical sessions emphasized research efforts in areas of technical performance of waste forms, processing aspects and modeling efforts of waste forms, and disposal systems. A special session addressed the technical aspects of managing and solidifying the waste currently stored in West Valley, New York.

Willrich, Mason, and others. **Radioactive Waste Management and Regulation.** New York: The Free Press, 1977. 138p. ISBN 0-02-934560-X.

This book focuses on the management and regulation of postfission radioactive waste, such as high-level and low-level waste contaminated with transuranic elements. It should be valuable in developing public policy and institutions necessary for safe management of radioactive waste.

Articles and Government Documents

Radioactivity

Environmental Aspects

Bondietti, E. A., F. O. Hoffman, and I. L. Larsen. "Air-to-Vegetation Transfer Rates of Natural Submicron Aerosols." *Journal of Environmental Radioactivity* 1 (1984): 5–27.

Cothern, C. Richard, William L. Lappenbusch, and Jacqueline Michel. "Drinking-Water Contribution to Natural Background Radiation." *Health Physics* 50 (Janaury 1986): 33–47.

Fairobent, James E., Edward F. Branagan, and Frank J. Congel. "Methodology for Calculating Doses from Radioactivity in the Environment." In Shyamal K. Majumdar and E. Willard Miller, eds., *Management of Radioactive Materials and Wastes: Issues and Progress* (Easton, PA: The Pennsylavania Academy of Science, 1985), 268–283. ISBN 0-9606670-4-0.

Green, B. M. R., and others. "Surveys of Natural Radiation Exposure in UK Dwellings with Passive and Active Measurement Techniques." *Science of the Total Environment* 45 (October 1985): 459–466.

Jester, William A., and Charley Yu. "Environmental Monitoring of Low-Level Radioactive Materials." In Shyamal K. Majumdar and E. Willard Miller, eds., *Management of Radioactive Materials and Wastes: Issues and Progress* (Easton, PA: The Pennsylvania Academy of Science, 1985), 294–311. ISBN 0-9606670-4-0.

Kates, R. W. "The Human Environment: The Road Not Taken, the Road Still Beckoning." *Annals of the Association of American Geographers* 77 (1987): 525–534.

Krumhansl, J. L. "Near Surface Heater Test Results: Environmental Implications for the Disposal of High Level Nuclear Waste." *Radioactive Waste Management and the Nuclear Fuel Cycle* 4 (September 1983): 1–31.

Miller, E. Willard, and Shyamal K. Majumdar. "Environmental and Biological Effects of Ionizing Radiation." In Shyamal K. Majumdar and E. Willard Miller, eds., *Management of Radioactive Materials and Wastes: Issues and Progress* (Easton, PA: The Pennsylvania Academy of Science, 1985), 342–355. ISBN 0-9606670-4-0.

Pearce, F. "Not in My Backyard!" *New Scientist* 100 (November 3, 1983): 346–351.

Rickard, W. H., and L. J. Kirby. "Trees as Indicators of Subterranean Water Flow from a Retired Radioactive Waste Disposal Site." *Health Physics* 52 (February 1987): 201–206.

Health Effects

Arthur, W. J., and others. "Radionuclide Export by Deer Mice at a Solid Radioactive Waste Disposal Area in Southeastern Idaho." *Health Physics* 52 (January 1987): 42–53.

Chau, Nguyen Phong. "Radiation Carcinogenesis in Humans: Is It Necessary to Revise Exposure Dose Limits Based on Recent Estimates of Lifetime Risks?" *Health Physics* 52 (June 1987): 753–761.

Cohen, Bernard L. "Alternatives to the BEIR Relative Risk Model for Explaining Atomic Bomb Survivor Cancer Mortality." *Health Physics* 52 (January 1987): 55–63.

______. "A Generic Probabilistic Risk Analysis for a High-Level Waste Repository." *Health Physics* 51 (October 1986): 519–528.

Danali, S., G. Margomenou, and K. Veldeki. "The Radioactivity of Spas on the Greek Island Ikaria and Influencing Factors." *Health Physics* 50 (April 1986): 509–513.

Evdokimoff, Victor. "Dose Assessment from Incineration of Deregulated Biomedical Radwaste." *Health Physics* 52 (March 1987): 325–329.

Fisher, Sidney T. "The Hazards of Nuclear Power Generation: The Case for Survival." *Long Range Planning* 16 (February 1983): 77–83.

Gilbert, Ethel S., Gerald R. Petersen, and Jeffrey A. Buchanan. "Mortality of Workers at the Hanford Site: 1945–1981." *Health Physics* 56 (January 1989): 11–25.

Haaland, Carsten M. "Forecasting Radiation Rates and Exposure from Multi-Aged Fallout." *Health Physics* 53 (December 1987): 613–622.

Hahn, N. A., Jr. "Disposal of Radium Removed from Drinking Water." *American Water Works Association Journal* 80 (July 1988): 71–78.

von Hippel, F. N., and others. "Civilian Casualties from Counterforce Attacks." *Scientific American* 259 (September 1988): 36–42.

Imahori, Akira. "Occupational Radiation Exposures During Maintenance Activities at Nuclear Power Plants in Japan." *Health Physics* 52 (March 1987): 367–370.

Johnsrud, Judith H. "Food Irradiation: Its Environmental Threat, Its Toxic Connection." *Workbook (Southwest Research and Info Center)* 13 (April/June 1988): 47–58.

Jury, M. R., and M. Mulholland. "Coastal Dispersion Conditions Near the Southwestern Tip of Africa: A System for Evaluation and Prediction." *Health Physics* 54 (April 1988): 421–429.

Kobal, I., and A. Renier. "Radioactivity of the Atomic Spa at Podćetrtek, Slovenia, Yugoslavia." *Health Physics* 53 (September 1987): 307–310.

Koch, J., and J. Tadmor. "Radfood—A Dynamic Model for Radioactivity Transfer Through the Human Food Chain." *Health Physics* 50 (June 1986): 721–737.

Lively, R. S., and E. P. Ney. "Surface Radioactivity Resulting from the Deposition of ^{222}Rn Daughter Products. *Health Physics* 52 (April 1987): 411–415.

Lubenau, Joel O., and Donald A. Nussbaumer. "Radioactive Contamination of Manufactured Products." *Health Physics* 51 (October 1986): 409–425.

Miller, Kenneth L. "Derived Benefits Which Lead to Radioactive Waste." In Shyamal K. Majumdar and E. Willard Miller, eds., *Management of*

Radioactive Materials and Wastes: Issues and Progress (Easton, PA: The Pennsylvania Academy of Science, 1985), 312–325. ISBN 0-9606670-4-0.

Ohnuki, Toshihiko, and Tadao Tanaka. "Migration of Radionuclides Controlled by Several Different Migration Mechanisms Through a Sandy Soil Layer." *Health Physics* 56 (January 1989): 47–53.

"On Radiation Hazards and Public Policy." *World Policy Journal* 4 (Spring 1987): 354–361.

Partanen, J. P., and I. Savolainen. "Significance of Contaminated Food in Collective Dose After a Severe Reactor Accident." *Health Physics* 50 (February 1986): 209–216.

"Protecting and Preserving the Nation's Food." *National Food Review* (Spring 1986): 1–26.

Singh, Harwant. "Prevention of Food Spoilage by Radiation Processing." *Canadian Home Economics Journal* 37 (Winter 1987): 5–10.

Smith, Neal, and Michael Baram. "The Nuclear Regulatory Commission's Regulation of Radiation Hazards in the Workplace: Present Problems and New Approaches to Reproductive Health." *Ecology Law Quarterly* 13:4 (1987): 879–938.

Stellman, Jeanne Mager. "Protective Legislation, Ionizing Radiation and Health: A New Appraisal and International Survey." *Women and Health* 12:1 (1987): 105–125.

Steyer, Robert. "Irradiated Food: A Marketing Hot Potato; The FDA Has Approved Use of the Process to Delay Spoilage in Fruits and Vegetables, But the Big Food Companies Are Not Exactly Lining Up to Irradiate." *Across the Board* 23 (July-August 1986): 14–21.

Stuchly, M. A. "Proposed Revision of the Canadian Recommendations on Radiofrequency-Exposure Protection." *Health Physics* 53 (December 1987): 649–665.

Sullivan, R. E., and Pao-Shan Weng. "Comparison of Risk Estimates Using Life-Table Methods." *Health Physics* 53 (August 1987): 123–134.

Swick, Craig. "FDA's Planning for Radiological Emergencies [U.S. Food and Drug Administration Guidelines for the Control of Radiation Contaminated Food]." *Journal of Energy Law & Policy* 2:1 (1981): 123–142.

Tadmor, Jacob. "Sensitivity Analysis of the Influence of Source-Term and Environmental Parameters on the Radiological Risk of Coal-Fired Plants. *Health Physics* 51 (July 1986): 61–80.

Turcotte, R. P. "Radiation Effects in High-Level Radioactive Waste Forms." *Radioactive Waste Management* 2 (December 1981): 169–177.

U.S. Congress. House. Committee on Interior and Insular Affairs. *Health Effects of Transmission Lines: Oversight Hearing, October 6, 1987.* 100th Cong., 1st sess., Washington, DC: GPO, 1988. 393p.

U.S. Congress. Senate. Committee on Energy and Natural Resources. Subcommittee on Energy Research and Development. *Department of Energy's Facilities for Defense Materials Production: Hearing, July 17, 1987: On the Public Health and Safety and Environmental Aspects of Operation of the Department of Energy's Facilities for Defense Materials Production.* 100th Cong., 1st sess., Washington, DC: GPO, 1987. 685p.

U.S. Congress. Senate. Committee on Environment and Public Works. Subcommittee on Nuclear Regulation. *Radiological Emergency Planning and Preparedness: Hearing, April 27, 1981.* 97th Cong., 1st sess., Washington, DC: GPO, 1981. 217p.

U.S. Congress. Senate. Committee on Veterans' Affairs. *VA Compensation and Other Service-Connected Benefits: Hearing, June 30, 1987, on S.9 [and Other Bills].* 100th Cong., 1st sess., Washington, DC: GPO, 1988. 401p.

Vaughan, Burton E. "Critical Considerations in the Assessment of Health and Environment Risks—What We Have Learned from the Nuclear Experience." *Science of the Total Environment* 28 (June 1983): 505–514.

Whicker, F. Ward, and T. B. Kirchner. "Pathway: A Dynamic Food-Chain Model to Predict Radionuclide Ingestion after Fallout Deposition." *Health Physics* 52 (June 1987): 717–737.

Zeighami, Elaine A., and Max D. Morris. "Thyroid Cancer Risk in the Population Around the Nevada Test Site." *Health Physics* 50 (January 1986): 19–32.

Radioactive Materials

Kawamura, Hisao. "Plutonium and Am Contamination of Tourist Property and Estimated Inhalation Intake of Visitors to Kiev after the Chernobyl Accident." *Health Physics* 52 (June 1987): 793–795.

La Duque, Winona. "Native America: The Economics of Radioactive Colonization [Exploitation of Uranium Resources on Navajo Land and the Laguna Pueblo, New Mexico]." *Review of Radical Political Economics* 15 (Fall 1983): 9–19.

Marshall, Walter. "Fast Reactors 'Some Questions and Answers.'" *Atom (Gt. Brit.)* (September 1980): 222–230.

______. "The Use of Plutonium [How Plutonium Could Be Used as a Vital Energy Source Producing the Maximum Economic Benefit with Minimum Proliferation Risk]." *Atom (Gt. Brit.)* (April 1980): 88–103.

Michel, Jacqueline, and C. Richard Cothern. "Predicting the Occurrence of ^{222}Ra in Ground Water." *Health Physics* 51 (December 1986): 715–721.

Organization for Economic Co-operation and Development. *Uranium: Resources, Production, and Demand: A Joint Report of the OECD Nuclear Energy Agency and the International Atomic Energy Agency.* Paris: OECD, 1986. 413p. ISBN 92-64-12842-5.

U.S. Bureau of Mines. *Availability of Rare-Earth, Yttrium, and Related Thorium Oxides: Market Economy Countries: A Minerals Availability Appraisal.* Washington, DC: GPO, 1986. 19p.

U.S. Congress. House. Committee on Armed Services. Procurement and Military Nuclear Systems Subcommittee. *Uranium Ore Residues: Potential Hazards and Disposition: Hearings, June 24–25, 1981.* 9th Cong., 1st sess., Washington, DC: GPO, 1981. 545p.

U.S. Congress. House. Committee on Government Operations. Environment, Energy, and Natural Resources Subcommittee. *Review of Hazardous Chemical Regulation at Nuclear Facilities by the Nuclear Regulatory Commission and Other Federal Agencies: Hearing, March 14, 1986.* 99th Cong., 2d sess., Washington, DC: GPO, 1986. 245p.

U.S. Congress. House. Committee on Interior and Insular Affairs. Subcommittee on Oversight and Investigations. *Military Plutonium from Commercial Spent Nuclear Fuel: Oversight Hearing, October 1, 1981, on Proposals to Obtain Plutonium from Commercial Spent Fuel for U.S. Nuclear Weapons.* 97th Cong., 1st sess., Washington, DC: GPO, 1982. 139p.

U.S. Congress. Senate. Committee on Energy and Natural Resources. Subcommittee on Energy Research and Development. *Uranium Revitalization, Tailings Reclamation, and Uranium Enrichment Programs: Hearing, April 10, 1986, on S 1004, a Bill to Authorize and Direct the Secretary of Energy to Establish a Program to Provide for Reclamation and Other Remedial Actions with Respect to Mill Tailings at Active Uranium and Thorium Processing Sites.* 99th Cong., 2d sess., Washington, DC: GPO, 1986. 415p.

Radon

Ackers, J. G., and others. "Radioactivity and Radon Exhalation Rates of Building Materials in the Netherlands." *Science of the Total Environment* 45 (October 1985): 151–156.

Alter, H. W. and R. L. Fleischer. "Passive Integrating Radon Monitor for Environmental Monitoring." *Health Physics* 40 (May 1981): 693–702.

Archibald, J. F., and H. J. Hackwood. "Radon Barriers for Underground Uranium Mine Use." *CIM Bulletin* 80 (December 1987): 39–42.

Brambley, M. R., and M. Gorfien. "Radon and Lung Cancer: Incremental Risks Associated with Residential Weatherization." *Energy* 11 (June 1986): 589–605.

Castren, O., and others. "Studies of High Indoor Radon Areas in Finland." *Science of the Total Environment* 45 (October 1985): 311–318.

Chittaporn, P., M. Eisenbud, and N. H. Harley. "A Continuous Monitor for the Measurement of Environmental Radon." *Health Physics* 41 (August 1981): 405–410.

Cohen, Bernard L. "A National Survey of ^{222}Rn in U.S. Homes and Correlating Factors." *Health Physics* 51 (August 1986): 175–183.

———, and Ernest S. Cohen. "Theory and Practice of Radon Monitoring with Charcoal Adsorption." *Health Physics* 45 (August 1983): 501–508.

———, and N. Gromicko. "Variation of Radon Levels in U.S. Homes with Various Factors." *JAPCA* 38 (February 1986): 129–134.

Dixon, K. L., and R. G. Lee. "Occurrence of Radon in Well Supplies." *American Water Works Association Journal* 80 (July 1988): 65–70.

Fleischer, R., and L. Turner. "Indoor Radon Measurement in the New York Capital District." *Health Physics* 46 (May 1984): 999–1011.

Fukui, M. "Soil Water Effects on Concentration Profiles and Variations of ^{222}Rn in a Vadose Zone." *Health Physics* 53 (August 1987): 181–186.

George, Andreas C. "Passive, Integrated Measurement of Indoor Radon Using Activated Carbon." *Health Physics* 46 (April 1984): 867–872.

Hagberg, N. "Some Tests on Measuring Methods for Indoor Radon Using Activated Charcoal." *Science of the Total Environment* 45 (October 1985): 417–423.

Harley, Naomi, and Bernard S. Pasternack. "Environmental Radon Daughter Alpha Dose Factors in a Five-Lobed Human Lung." *Health Physics* 42 (June 1982): 789–799.

———. "A Model for Predicting Lung Cancer Risks Induced by Environmental Levels of Radon Daughters." *Health Physics* 40 (March 1981): 307–316.

Hart, Kaye P., and Desmond M. Levins. "Steady-Site Rn Diffusion Through Tailings and Multiple Layers of Covering Materials." *Health Physics* 50 (February 1986): 369–379.

Jacobs, W., and H. G. Paretzke. "Risk Assessment for Indoor Exposure to Radon Daughters." *Science of the Total Environment* 45 (October 1985): 551-562.

Kerr, R. A. "Indoor Radon: The Deadliest Pollutant." *Science* (April 29, 1988): 606-608.

Longtin, J. P. "Occurrence of Radon, Radium and Uranium in Groundwater." *American Water Works Association Journal* 80 (July 1988): 84-93.

Lowry, J. D., and S. B. Lowry. "Radionuclides in Drinking Water." *American Water Works Association Journal* 80 (July 1988): 50-64.

Myrick, T. E., B. A. Berven, and F. F. Haywood. "Determination of Concentrations of Selected Radionuclides in Surface Soil in the U.S." *Health Physics* 45 (September 1983): 631-642.

Nazaroff, W. W. "Optimizing the Total-Alpha Three-Count Technique for Measuring Concentrations of Radon Progeny in Residences." *Health Physics* 46 (February 1984): 395-405.

______. "Radon Transport into a Detached One-story House with a Basement." *Atmospheric Environment* 19:1 (1985): 31-46. Discussion 20:5 (1986): 1065-1067.

______, and others. "Potable Water as a Source of Airborne ^{222}Rn in U.S. Dwellings: A Review and Assessment." *Health Physics* 52 (March 1987): 281-295.

Nero, A. V., and W. W. Nazaroff. "Characterizing the Source of Radon Indoors." *Radiation Protection Dosimetry* 7 (1984): 23-29.

Nyberg, P., and D. Bernhardt. "Measurement of Time Integrated Radon Concentrations in Residences." *Health Physics* 45 (August 1983): 539-543.

Paredes, C. H., and others. "Radionuclide Content of and ^{222}Rn Emanation from Building Materials Made from Phosphate Industry Waste Products." *Health Physics* 53 (July 1987): 23-29.

Paschoa, A. S., J. A. Torrey, and M. E. Wrenn. "Reducing Radon Exhalation from Covered Tailings: Optimization or Cost Effectiveness?" *Science of the Total Environment* 45 (October 1985): 187-194.

Poffijn, A., and others. "Results of a Preliminary Survey on Radon in Belgium." *Science of the Total Environment* 45 (October 1985): 335-342.

Proposed Standard for Radon-222 Emissions from Licensed Uranium Mill Tailings, Draft Economic Analysis. Prepared by Jack Faucett Associates. Washington, DC: U.S. Environmental Protection Agency, Office of Radiation Programs, 1986. 199p.

Put, L. W., R. J. DeMeijer, and B. Hogeweg. "Survey of Radon Concentrations in Dutch Dwellings." *Science of the Toal Environment* 45 (October 1985): 441-448.

"Radon: Pinpointing a Mystery," *EPA (Environmental Protection Agency) Journal* 12 (August 1986): 2-15.

Radon Reduction Techniques for Detached Houses, Technical Guidance. Research Triangle Park, NC: Air and Energy Engineering Research Laboratory, Office of Environmental Engineering and Technology, Office of Research and Development, U.S. Environmental Protection Agency, 1986. 50p.

Raes, F., A. Janssens, and H. Vanmarcke. "A Closer Look at the Behavior of Radioactive Decay Products in Air." *Science of the Total Environment* 45 (October 1985): 205-218.

Reineking, A., K. H. Becker, and J. Porstendörfer. "Measurements of the Unattached Fractions of Radon Daughters in Houses." *Science of the Total Environment* 45 (October 1985): 261-270.

Ronca, Battista, and others. Interim Indoor Radon and Radon Decay Product Measurements Protocols. Washington, DC: U.S. Environmental Protection Agency, Office of Radiation Programs, 1986.

Rudnick, S. N., and E. F. Maher. "Surface Deposition of ^{222}Rn Decay Product With and Without Enhanced Air Motion." *Health Physics* 51 (September 1986): 283-293.

Schmier, H., and A. Wicke. "Results from a Survey of Indoor Radon Exposures in the Federal Republic of Germany. *Science of the Total Environment* 45 (October 1985): 307-310.

Steinhäusler, F., and others. "Radiation Exposure of the Respiratory Tract and Associated Carcinogenic Risk Due to Inhaled Radon Daughters." *Health Physics* 45 (August 1983): 331-337.

Stranden, E. "Thoron and Radon Daughters in Different Atmospheres." *Health Physics* 38 (May 1980): 777-785.

U.S. Congress. House. Committee on Energy and Commerce. Subcommittee on Transportation, Tourism, and Hazardous Materials. *Radon Pollution Control Act of 1987: Hearing, April 23, 1987.* 100th Cong., 1st sess., Washington, DC: GPO, 1987. 119p.

U.S. Congress. House. Committee on Science and Technology. *Radon and Indoor Air Pollution: Hearing, October 10, 1985, before the Subcommittee on Natural Resources, Agriculture Research and Environment.* 99th Cong., 1st sess., Washington, DC: GPO, 1986. 291p.

U.S. Congress. House. Committee on Science and Technology. Subcommittee on Natural Resources, Agriculture Research and Environment. *Radon and Indoor Air Pollution: Hearing, October 10, 1985.* 99th Cong., 1st sess., Washington, DC: GPO, 1986. 291p.

______. *Residential Radon Contamination and Indoor Quality Research Needs: Hearing, September 17, 1986.* 99th Cong., 2nd sess., Washington, DC: GPO, 1987. 333p.

U.S. Congress. Senate. Committee on Environment and Public Works. *Indoor Air Pollution: Hearing, August 5, 1985, on S.1198, a Bill to Establish in the Environmental Protection Agency a Program of Research on Indoor Air Quality and for Other Purposes.* 99th Cong., 1st sess., Washington, DC: GPO, 1985. 68p.

______. *Radon Gas Issues: Joint Hearings, April 2, 1987, on S.743 and S.744, before the Subcommittee on Environmental Protection and Superfund and Environmental Oversight.* 100th Cong., 1st sess., Washington, DC: GPO, 1987. 124p.

U.S. Congress. Senate. Committee on Environment and Public Works. Subcommittee on Environmental Protection. *Health Effects of Indoor Air Pollution: Hearing, April 24, 1987.* 100th Cong., 1st sess., Washington, DC: GPO, 1987. 124p.

Van der Lugt, G., and L. C. Scholten. "Radon Emanation from Concrete and the Influence of Using Fly Ash in Cement." *Science of the Total Environment* 45 (October 1985): 143–150.

Vanmarcke, H., A. Janssens, and F. Raes. "The Equilibrium of Attached and Unattached Radon Daughters in the Domestic Environment." *Science of the Total Environment* 45 (October 1985): 251–260.

Wilkening, Marvin, and Andreas Wicke. "Seasonal Variation of Indoor Rn at a Location in the Southwestern United States." *Health Physics* 51 (October 1986): 427–436.

Wilkinson, P., and P. J. Dimbylon. "Radon Diffusion Modelling." *Science of the Total Environment* 45 (October 1985): 227–232.

Wilkniss, P. E., and R. E. Larson. "Atmospheric Radon Measurements in the Arctic: Fronts, Seasonal Observations, and Transport of Continental Air to Polar Regions." *Journal of the Atmospheric Sciences* 41 (August 1, 1984): 2347–2358.

Zarcone, M. J., and others. "A Comparison of Measurements of Thoron, Radon and Their Daughters in a Test House with Model Predictions." *Atmospheric Environment* 20:6 (1986): 1273–1279.

Nuclear Warfare

Antinuclear Movement

Akaho, Tsuneo. "Japan's Three Nonnuclear Principles: A Coming Demise?" *Peace and Change* 11 (Spring 1985): 75–89.

Byrne, Paul. "The Campaign for Nuclear Disarmament: The Resilience of a Protest Group." *Parliamentary Affairs (London)* 40 (October 1987): 517-535.

Carney, James L. "Is It Ever Moral To Push the Button?" *Parameters* 18 (March 1988): 73-87.

Chafer, Tony. "Politics and the Perception of Risk: A Study of the Anti-Nuclear Movements in Britain and France." *West European Politics* 8 (January 1985): 5-23.

Downey, Gary L. "Ideology and the Clamshell Identity: Organizational Dilemmas in the Anti-Nuclear Power Movement." *Social Problems (Social Study Social Problems)* 23 (June 1986): 357-373.

Flavin, Christopher. "Chernobyl: The Political Fallout in Western Europe." *Forum for Applied Research and Public Policy* 2 (Summer 1987): 16-28.

Garfinkle, Adam M. "The Passions of the Nuclear Freeze Movement." *Crossroads* no. 21 (1986): 83-108.

______. "The Unmaking of the Nuclear Freeze." *Washington Quarterly* 8 (Spring 1985): 109-120.

Joffe, Josef. "Peace and Populism: Why the European Anti-Nuclear Movement Failed." *International Security* 11 (Spring 1987): 3-40.

Johansen, Robert. "Global Security Without Nuclear Deterrence." *Alternatives (Centre Study Developing Societies)* 12 (October 1987): 435-460.

Karpeles, Michael D. "Congressional Nuclear Freeze Proposals: Constitutionality and Enforcement.: *Harvard Journal on Legislation* 23 (Summer 1986): 483-557.

Kitschelt, Herbert P. "Political Opportunity Structures and Political Protest: Anti-Nuclear Movements in Four Democracies." *British Journal of Political Science* 16 (January 1986): 57-85.

Klandermans, Bert, and Dirk Oegema. "Campaigning for a Nuclear Freeze: Grass Roots Strategies and Local Governments in the Netherlands." In *Research in Political Sociology, 1987,* edited by Richard G. Braungart. Greenwich, CT: JAI Press, 1987, pp. 305-337.

Lamare, James W. "International Conflict: ANZUS and New Zealand Public Opinion." *Journal of Conflict and Resolution* 31 (September 1987): 420-437.

Lange, David. "Nuclear Policy Sparks Debate." *New Zealand Foreign Affairs Review* 35 (January-March 1985): 3-17.

Mack, Andrew. "Crisis in the Other Alliance: ANZUS in the 1980's." *World Policy Journal* 3 (Summer 1986): 447–472.

Messina, Anthony M. "Postwar Protest Movements in Britain: A Challenge to Parties." *Review of Politics* 49 (Summer 1987): 410–428.

Nusbaumer, Michael R., and Judith A. Dilorio. "The Medicalization of Nuclear Disarmament Claims." *Peace and Change* 11 (Spring 1985): 63–73.

Payne, Keith B., and Jill E. Coleman. "Christian Nuclear Pacifism and Just War Theory: Are They Compatible?" *Comparative Strategy* 7:1 (1988): 75–89.

Pugh, Michael. "Wellington Against Washington: Steps to Unilateral Arms Control." *Arms Control (London)* 7 (May 1986): 63–73.

Rizzo, Robert. "Nuclear Warfare: The Psychological Effects and Their Impact on Moral Reasoning." *International Journal of Social Economics* 13:1-2 (1986): 40–54.

"Special Issue on Peace Movements." *Journal of Peace Research* 23 (June 1986): 97–208.

Takahara, Takao. "Local Government Initiatives to Promote Peace." *Peace and Change* 12:3–4 (1987): 51–58.

Turner, Royce Logan. "Political Opposition to Nuclear Power: An Overview." *Political Quarterly* 57 (October–December 1986): 438–443.

Willoughby, M. J. "Nuclear Freeze Attitude Structures: A Pre- and Post-Test of the Impact of the Korean Airliner Incident." *Social Science Quarterly* 67 (September 1986): 534–544.

Zhukov, Yu. "The Present-Day International Situation and the Peace Movement." *International Affairs (Moscow)* (July 1985): 36–44.

Effects of Atomic Bombing

Altfeld, Michael F., and Stephen J. Cimbala. "Targeting for Nuclear Winter: A Speculative Essay." *Parameters* 15 (Autumn 1985): 8–15.

Beck, H. E., and P. W. Krey. Radiation Exposures in Utah from Nevada Nuclear Tests." *Science* 220 (April 1, 1983): 18–24.

Bower, Peter. "After the Bomb [Effects of a Nuclear Blast on an Urban Center]." *Political Affairs* 61 (May 1982): 15–23.

Broyles, A. A. "Worldwide Radioactivity from a Nuclear War." *American Journal of Physics* 54 (February 1986): 151–157.

Ehrlich, Paul R. "The Nuclear Winter: Discovering the Ecology of Nuclear War." *Amicus Journal* 5 (Winter 1984): 20–30.

Ehrlich, Robert. "We Should Not Overstate the Effects of Nuclear War." *International Journal of World Peace* 3 (July–September 1986): 31–46.

Gouré, Leon. "'Nuclear Winter' in Soviet Mirrors." *Strategic Review* 13 (Summer 1985): 22–38.

Hancock, Don. "The Wasting of America: Target—Nevada, Target—New Mexico." *Workbook (Southwest Research and Info Center)* 13 (Jan–March 1988): 2–12.

Hoeber, Francis P., and Robert K. Squire. "The 'Nuclear Winter' Hypothesis: Some Policy Implications." *Strategic Review* 13 (Summer 1985): 39–46.

Horowitz, Dan, and Robert J. Leiber. "Nuclear Winter and the Future of Deterrence." *Washington Quarterly* 8 (Summer 1985): 59–70.

Loewe, W. E. "Hiroshima and Nagasaki Radiations: Delayed Neutron Contributions and Comparison of Calculated and Measured Cobalt Activations." *Nuclear Technology* 63 (March 1985): 311–318.

______. "Initial Radiations from Tactical Nuclear Weapons." *Nuclear Technology* 70 (August 1985): 274–284.

Mason, Leonard. "The Bikinians: A Transplanted Population." *Human Organization* (Spring 1950): 5–15.

Payne, B. A., and others. "The Effects on Property Values of Proximity to a Site Contaminated with Radioactive Waste." *Natural Resources Journal* 27 (Summer 1987): 579–590.

Sagan, Carl. "Nuclear War and Climatic Catastrophe: Some Policy Implications." *Foreign Affairs* 62 (Winter 1983–1984): 257–292.

Shenfield, Stephen. "Nuclear Winter and the USSR." *Millennium* 15 (Summer 1986): 197–208.

Thompson, Starley L., and Stephen H. Schneider. "Nuclear Winter Reappraised." *Foreign Affairs* 64 (Summer 1986): 981–1005.

Titus, A. Costandina. "Governmental Responsibility for Victims of Atomic Testing: A Chronicle of the Politics of Compensation." *Journal of Health Politics, Policy and Law* 8 (Summer 1983): 277–292.

U.S. Congress. House. Committee on Interstate and Foreign Commerce, Subcommittee on Oversight and Investigations. *"The Forgotten Guinea Pigs": A Report on Health Effects of Low-Level Radiation Sustained as a Result of the Nuclear Weapons Testing Program Conducted by the United States Government; Report, August 1980.* 96th Cong., 2d sess., Washington, DC: GPO, 1980. 42p.

———. *Low-Level Radiation Effects on Health: Hearings, April 23–August 1, 1979.* 96th Cong., 1st sess., Washington, DC: GPO, 1979. 1188p.

U.S. Congress. House. Committee on Science and Technology. *Nuclear Winter: Joint Hearing, March 14, 1985, Before the Subcommittee on Natural Resources, Agriculture Research and Environment of the Committee on Science and Technology and the Subcommittee on Science and the Environment of the Committee on Interior and Insular Affairs.* 99th Cong., 1st sess., Washington, DC: GPO, 1985. 313p.

U.S. Congress. House. Committee on Science and Technology, Subcommittee on Investigations and Oversight. *The Consequences of Nuclear War on the Global Environment: Hearing, September 15, 1982.* 97th Cong., 2d sess., Washington, DC: GPO, 1983. 260p.

U.S. Congress. House. Committee on Veterans' Affairs, Subcommittee on Oversight and Investigations. *Review of Federal Studies on Health Effects of Low-Level Radiation Exposure and Implementation of Public Law 97-72: Hearing, May 24, 1983.* 98th Cong., 1st sess., Washington, DC: GPO, 1983. 472p.

U.S. Congress. Senate. Committee on the Judiciary. *Government Liability for Atomic Weapons Testing Program: Hearing, July 27, 1986, on S.2454 and on H.R.1338.* 99th Cong., 2d sess., Washington, DC: GPO, 1987. 163p.

U.S. Congress. Senate. Committee on Veterans' Affairs. *Veterans' Exposure to Ionizing Radiation as a Result of Detonations of Nuclear Devices: Hearing, April 6, 1983.* 98th Cong., 1st sess., Washington, DC: GPO, 1984. 577p.

Van Dyke, J., K. R. Smith, and S. Siwatibau. "Nuclear Activities and the Pacific Islanders." *Energy* 9 (1984): 733–750.

Weisgall, Jonathan M. "The Nuclear Nomads of Bikini." *Foreign Policy* (Summer 1980): 74–98.

Nuclear Nonproliferation

Agyeman-Duah, Baffour. "Nuclear Weapons Free Zones and Disarmament." *Africa Today* 32 (First/Second Quarters 1985): 77–89.

Alves, Dora. "New Zealand and ANZUS: An American View." *Round Table* (April 1987): 207–222.

Borisov, K. "The Effectiveness of the Nuclear Non-proliferation Regime." *International Affairs (Moscow)* (December 1982): 101–106.

Clausen, Peter A. "Nonproliferation Illusions: Tarapur in Retrospect [U.S. Peaceful Nuclear Cooperation in India]." *Orbis* 27 (Fall 1983): 741–759.

Donnelly, Warren H. "Nonproliferation Policy of the United States in the 1980s." *SAIS (School Advanced International Studies) Review* 7 (Summer-Fall 1987): 159–179.

Dorian, Thomas F., and Leonard S. Spector. "Covert Nuclear Trade and the International Nonproliferation Regime [Dealings Since the Mid-1960s]." *Journal of International Affairs* 35 (Spring–Summer 1981): 29–68.

"Europe's Peace Movement: Time Bomb for U.S.? [Growing Backlash Against a North Atlantic Treaty Organization Strategy That Calls for the Deployment of Nuclear Weapons in Western Europe]." *U.S. News* 91 (November 9, 1981): 23–54.

Forkner, Claude E. "A New Plan for the Elimination of All Atomic Weapons: A Referendum for Human Destiny." *International Journal on World Peace* 2 (July–September 1985): 63–89.

Goldblat, Josef. "The Third Review Conference of the Nuclear Nonproliferation Treaty." *Bulletin of Peace Proposals* 17:1 (1986): 13–27.

Graham, Kennedy. "New Zealand's Non-Nuclear Policy: Towards Global Security." *Alternatives (Centre Study Developing Societies)* 12 (April 1987): 217–242.

Gummett, Philip. "From NPT to INFCE: Developments in Thinking About Nuclear Non-proliferation [the Nuclear Nonproliferation Treaty and the International Nuclear Fuel Cycle Evaluation] *International Affairs (London)* 57 (Autumn 1981): 549–567.

Hall, B. Welling. "The Antinuclear Peace Movement: Toward an Evaluation of Effectiveness." *Alternatives (World Policy Institute)* 9 (Spring 1984): 475–517.

Hanson, F. Allan. "Trouble in the Family: New Zealand's Antinuclear Policy." *SAIS (School Advanced International Studies) Review* 7 (Winter-Spring 1987): 139–155.

Johnson-Fresse, Joan. "Interpretations of the Nonproliferation Treaty: The U.S. and West Germany [and the Interpretation of Article IV, which Grants

Access to Nuclear Energy for Peaceful Purposes to Nonnuclear-Weapon States]." *Journal of International Affairs* 37 (Winter 1984): 283-293.

Kennedy, Richard T. "The Nonproliferation Predicament [U.S. Commitment to the Objective of Preventing the Spread of Nuclear Weapons]." *Society* 20 (September-October 1983): 30-60.

______. "Nonproliferation: Where We Are and Where We're Going." *Department of State Bulletin* 83 (December 1983): 52-57.

Khilnani, Niranjan M. "The Denuclearization of South Asia." *Round Table* (July 1986): 280-286.

Klick, Donna J. "A Balkan Nuclear Weapon-Free Zone: Viability of the Regime and Implications for Crisis Management." *Journal of Peace Research* 24 (June 1987): 111-124.

Lippman, Matthew. "The South Pacific Nuclear Free Zone Treaty: Regional Autonomy Versus International Law and Politics." *Loyola of Los Angeles International and Comparative Law Journal* 10:1 (1988): 109-133.

Lovins, Amory B., and others. "Nuclear Power and the Nuclear Bombs [and the Nonproliferation Problem]." *Foreign Affairs* 58 (Summer 1980): 1137-1177.

O'Brien, John N. "International Auspices for the Storage of Spent Nuclear Fuel as a Nonproliferation Measure." *Natural Resources Journal* 21 (October 1981): 857-894.

Milhollin, Gary. "Heavy Water Cheaters." *Foreign Policy* (Winter 1987/1988): 100-119.

Multan, Wojciech. "The Past and Future of Nuclear-Weapon-Free Zones." *Bulletin of Peace Proposals* 16:4 (1985): 375-385.

Nusbaumer, Michael R., and Judith A. Dilorio. "The Mechanization of Nuclear Disarmament Claims." *Peace and Change* 11 (Spring 1985): 63-73.

Paul, T. V. "Nuclear-Free-Zone in the South Pacific: Rhetoric or Reality?" *Round Table* (July 1986): 252-262.

Payne, Keith B., and Jill E. Coleman. "Christian Nuclear Pacifism and Just War Theory: Are They Compatible?" *Comparative Strategy* 7:1 (1988): 75-89.

Poneman, Daniel. "Nuclear Policies in Developing Countries [and the Nuclear Nonproliferation Treaty]." *International Affairs (London)* 57 (Autumn 1981): 568-584.

Power, Paul F. "The Mixed State of Non-Proliferation: The NPT Review Conference and Beyond." *International Affairs (London)* 62 (Summer 1986): 477–491.

"Preventing the Spread of Nuclear Weapons." *Disarmament (UN)* 8 (Spring 1985): 1–107; 8 (Summer 1985): 131–145.

Redick, John R. "The Tlatelolco Regime and Nonproliferation in Latin America [the Latin American Nuclear-Weapon-Free Zone Established by the Treaty of Tlatelolco (Treaty for the Prohibition of Nuclear Weapons in Latin America) in 1969]." *International Organization* 35 (Winter 1981): 103–134.

Rhodes, Edward. "Nuclear Weapons and Credibility: Deterrence Theory Beyond Rationality." *Review of International Studies* 14 (January 1988): 45–62.

Sanders, Ben. "Non-proliferation in the year 2000." *Disarmament (UN)* 9 (Summer 1986): 93–107.

Schorr, Brian L. "Testing Statutory Criteria for Foreign Policy: The Nuclear Non-Proliferation Act of 1978 and the Export of Nuclear Fuel to India." *New York University Journal of International Law and Politics* 14 (Winter 1982): 419–466.

Smith, Roger K. "Explaining the Non-proliferation Regime: Anomalies for Contemporary International Relations Theory." *International Organization* 41 (Spring 1987): 253–281.

"Third Review Conference Held for Nonproliferation Treaty. *Department of State Bulletin* 85 (November 1985): 35–44.

Triggs, Gillian. "Australia's Bilateral Nuclear Safeguards Agreements: A Triumph of Reality Over Symbolism?" *Australian Outlook* 39 (December 1985): 147–155.

U.S. Congress. House. Committee on Foreign Affairs. *Amendments to the Foreign Assistance Act of 1961: Nuclear Prohibitions and Certain Human Rights Matters: Hearings and Markup, November 20 and December 8, 1981, on H.R. 5015; H. Res. 286.* 97th Cong., 1st sess., Washington, DC: GPO, 1982. 81p.

______. *Controls on Exports of Nuclear-Related Goods and Technology: Hearing, June 24, 1982, Before the Subcommittees on International Security and Scientific Affairs and on International Economic Policy and Trade.* 97th Cong., 2d sess., Washington, DC: GPO, 1983. 86p.

______. *Legislation to Amend the Nuclear Non-Proliferation Act of 1978: Hearings and Markup, August 3–December 14, 1982, Before the Subcommittees on International Security and Scientific Affairs and on International Economic Policy and Trade, on H.R. 6032 and H.R. 6318.* 97th Cong., 2d sess., Washington, DC: GPO, 1983. 456p.

U.S. Congress. House. Committee on Foreign Affairs. *Proposed Amendments to the Nuclear Non-Proliferation Act, 1983: Hearings, September 20–November 1, 1983, on H.R. 1417 and H.R. 3058.* 98th Cong., 1st sess., Washington, DC: GPO, 1984. 382p.

______. *Review of GAO [General Accounting Office] Report on the Nuclear Non-Proliferation Act of 1978: Hearing May 21, 1981, Before the Subcommittee on International Security and Scientific Affairs and on International Economic Policy and Trade.* 97th Cong., 1st sess., Washington, DC: GPO, 1981. 32p.

______. Subcommittee on Arms Control, Oceans, International Operations and Environment. *The Nonproliferation Treaty: Hearing, July 24, 1980.* 96th Cong., 2d sess., Washington, DC: GPO, 1980. 99p.

U.S. Congress. House. Committee on Interior and Insular Affairs. Subcommittee on Energy and the Environment. *United States Nuclear Nonproliferation Policy: Oversight Hearing: Pt. 10, July 26, 1979.* 96th Cong., 1st sess., Washington, DC: GPO, 1980. 125p.

U.S. Congress. House. Committee on Science and Technology. Subcommittee on Energy Research and Production. *Technical Aspects of Nuclear Nonproliferation: Safeguards: Hearings, August 3–4, 1982.* 97th Cong., 2d sess., Washington, DC: GPO, 1982. 852p.

U.S. Congress. Senate. Committee on Foreign Relations. *IAEA Programs for Safeguards: Hearing, December 2, 1981.* 97th Cong., 1st sess., Washington, DC: GPO, 1982. 104p.

______. *Prohibition of Nuclear Weapons in Latin America: Hearing, September 22, 1981, on Ex.I, 95-2, the Additional Protocol to the Treaty for the Prohibition of Nuclear Weapons in Latin America, Also Known as the Treaty of Tlatelolco.* 97th Cong., 1st sess., Washington, DC: GPO, 1981. 21p.

______. *U.S. Nuclear Nonproliferation Policy: Hearing, September 29, 1982.* 97th Cong., 2d sess., Washington, DC: GPO, 1983. 81p.

U.S. Congress. Senate. Committee on Governmental Affairs. *Nuclear Nonproliferation and U.S. National Security: Hearings, February 24–March 5, 1987.* 100th Cong., 1st sess., Washington, DC: GPO, 1987. 225p.

______. Subcommittee on Energy, Nuclear Proliferation and Governmental Processes. *Nuclear Nonproliferation Policy: Hearing, June 24, 1981.* 97th Cong., 1st sess., Washington, DC: GPO, 1981. 177p.

Vinogradov, Yuri. "The Indian Ocean: The Problem of Demilitarization." *International Affairs (Moscow)* (July 1987): 58–64.

Walters, Ronald. "The United States and South Africa: Nuclear Collaboration Under the Reagan Administration." *Trans-Africa Forum* 2 (Fall 1983): 17–30.

Nuclear Proliferation

Ausness, Richard. "Putting the Genie Back in the Bottle: U.S. Controls Over Sensitive Nuclear Technology." *George Washington Journal of International Law and Economics* 16:1 (1982): 65–118.

Blix, Hans. "Can International Safeguards Stop Nuclear Proliferation?" *Disarmament (UN)* 6 (Autumn/Winter 1987): 1–7.

Cutter, Susan L., and others. "From Grass Roots to Partisan Politics: Nuclear Freeze Referenda in New Jersey and South Dakota." *Political Geography Quarterly* 6 (October 1987): 287–300.

Goheen, Robert F. "Problems of Proliferation: U.S. Policy and the Third World." *World Politics (Princeton)* 35 (January 1983): 194–215.

Kelleher, Catherine M. "Western Europe: Cycles, Crisis, and the Nuclear Revolution." *Annals of the American Academy of Political and Social Science* 469 (September 1983): 91–103.

Khachaturov, Karen. "Moscow Forum and Its Significance." *International Affairs (Moscow)* (May 1987): 82–87.

Lange, David. "New Zealand's Security Policy." *Foreign Affairs* 63 (Summer 1985): 1009–1019.

______. "Relations With the U.S." *New Zealand Foreign Affairs* 35 (July-September 1985): 29–35.

Milhollin, Gary. "Dateline New Delhi: India's Nuclear Cover-up." *Foreign Policy* (Fall 1986): 161–175.

Poneman, Daniel. "Nuclear Proliferation Prospects for Argentina." *Orbis* 27 (Winter 1984): 853–880.

"Proliferation and the Pakistan Bomb." *Arms Control Today* 17 (November 1987): 3–17.

"Prospects of a Nuclear-Free Northern Europe." *World Marxist Review* 30 (March 1987): 103–111.

Quester, George H., ed. "Nuclear Proliferation: Breaking the Chain: A Special Issue." *International Organization* 35 (Winter 1981): 1–240.

Samuel, Peter, and F.P. Serong. "The Troubled Waters of ANZUS." *Strategic Review* 14 (Winter 1986): 39–48.

Shultz, George. "Preventing the Proliferation of Nuclear Weapons." *Department of State Bulletin* 84 (December 1984): 17–21.

Sowden, R. G. "Extra Dividends from Nuclear Technology." *Nuclear Energy* 24 (February 1985): 11–19.

"Special Nuclear Issue." *Crossroads* 21 (1986): 5–108.

Spector, Leonard S. "Silent Spread." *Foreign Policy* (Spring 1985): 53–78.

Sweet, William. "Controlling Nuclear Proliferation." *Editorial Research Reports* (July 17, 1981): 511–532.

Ter Haar, Bas, and Piet De Klerk. "Verification of Non-Production: Chemical Weapons and Nuclear Weapons Compared." *Arms Control (London)* 8 (December 1987): 197–212.

Titterton, Sir Ernest. "The Fallacies of Nuclear-Free Zoning and the Importance of Nuclear Energy [Argues for the Continued Use and Development of Nuclear Energy]." *International Journal on World Peace* 1 (Autumn 1984): 56–73.

U.S. Congress. *Nuclear Proliferation Factbook: Prepared for the Subcommittees on Arms Control, International Security and Science and on International Economic Policy and Trade of the Committee on Foreign Affairs, U.S. House of Representatives and the Subcommittee on Energy; Nuclear Proliferation and Federal Processes of the Committee on Governmental Affairs, U.S. Senate.* 99th Cong., 1st sess., Washington, DC: GPO, 1985. 591p.

U.S. Congress. House. Committee on Foreign Affairs. *Nuclear Proliferation: Dealing with Problem Countries: Hearing, July 23, 1981, Before the Subcommittees on International Security and Scientific Affairs and on International Economic Policy and Trade.* 97th Cong., 1st sess., Washington, DC: GPO, 1981. 57p.

______. *Proposed Nuclear Cooperation Agreement with the People's Republic of China, Hearing and Markup, July 31 and November 13, 1985, on H.J. Res 404*. 99th Cong., 1st sess., Washington, DC: GPO, 1987. 243p.

U.S. Congress. House. Committee on Science and Technology. Subcommittee on Energy Research and Production. *Nuclear Safeguards: A Reader; Report, December 1983*. Compiled and edited by Warren H. Donnelly and Joseph F. Pilat. 98th Cong., 1st sess., Washington, DC: GPO, 1983. 999p.

U.S. Congress. Senate. Committee on Foreign Relations. *United States-People's Republic of China Nuclear Agreement: Hearing, October 9, 1985*. 99th Cong., 1st sess., Washington, DC: GPO, 1986. 315p.

U.S. Congress. Senate. Committee on Governmental Affairs. *U.S. Policy on Export of Helium-3 and Other Nuclear Materials and Technology: Hearing May 13, 1982*. 97th Cong., 2d sess., Washington, DC: GPO, 1982. 153p.

Nuclear Energy

General Works

Fremlin, J. H. "Power Production at Minimum Risk [Compares Risks to the British Population from the Main Sources for Power and Concludes That No Power Risks Are Large-Enough To Be the Deciding Factor in Choosing the Best Source of Power]." *Nuclear Energy* 22 (April 1983): 67–73.

Loftus, J. E. "Economic Aspects of Atomic Power: An Economist Analyzes the Recent Study on Atomic Power Sponsored by the Cowles Commission." *Bulletin of the Atomic Scientists* 7 (March 1951): 70–74.

Lovins, Amory B. "Energy Strategy: The Road Not Taken." *Foreign Affairs* 55 (October 1976): 65–96.

Nuclear Power from Fission Reactors. Washington, DC: U.S. Department of Energy, 1982. 21p.

Pasqualetti, M. J. "The Dissemination of Geographical Findings on Nuclear Power." *Transactions of the Institute of British Geographers NS* 11 (1986): 325–336.

Pryde, P. R. "Nuclear Power." In L. Diens and T. Shabad, *The Soviet Energy System: Resource Use and Policies* (Washington, DC: V. H. Winston & Sons, 1979), 151–170.

Solomon, B. D. "The Impact of Nuclear Power Plant Dismantlement on Radioactive Waste Disposal." *Man, Environment, Space, and Time* 4 (1984): 39–60.

Temples, James R. "The Politics of Nuclear Power: A Subgovernment in Transition." *Political Science Quarterly* 95 (Summer 1980): 239–260.

Accidents

Ambach, W., and others. "Chernobyl Fallout on Alpine Glaciers." *Health Physics* 56 (January 1989): 27–31.

ApSimon, H. M., and others. "The Use of Weather Radar in Assessing Deposition of Radioactivity from Chernobyl Across England and Wales." *Atmospheric Environment* 22:9 (1988): 1895–1900.

Ballestra, S. B., and others. "Fallout Deposition at Monaco Following the Chernobyl Accident." *Journal of Environmental Radioactivity* 5 (1987): 391–400.

Berlow, Alan. "Three Mile Reactor Accident Clouds Future of Industry: Nuclear Power Crossroads [Effects of the Accident at a Nuclear Plant Near Harrisburg, PA., March 28, 1979]." *Congressional Quarterly Weekly Report* 37 (April 7, 1979): 627–630.

"Chernobyl: Containing the Damage." *Current Digest of the Soviet Press* 38 (June 18, 1986): 11–17.

Corvisiero, P., and others. "Radioactivity Measurements in Northwest Italy After Fallout from the Reactor Accident at Chernobyl." *Health Physics* 53 (July 1987): 83–87.

Duffy, L. P., and others. "The Three Mile Accident and Recovery." *Nuclear Energy* 25 (August 1986): 199–215.

Edsall, John T. "Hazards of Nuclear Fission Power and the Choice of Alternatives." *Environmental Conservation* 1 (1974): 21–30.

Faltermayer, Edmund. "Nuclear Power After Three Mile Island: Even Before the Pennsylvania Accident Raised New Doubts About the Safety of Fission Plants, Their Economic Advantage Was Narrowing and the Utilities Were Turning Back to Coal." *Fortune* 99 (May 7, 1979): 114–118.

Ferrar, Terry A. "Three Mile Island—The Regulatory Challenge of 1979 [Distribution Costs of the Atomic Power Plant Accident Near Harrisburg, PA]." *Public Utilities Fortnightly* 104 (July 19, 1979): 15–18.

Fetter, S. A., and K. Tsipis. "Catastrophic Releases of Radioactivity." *Scientific American* 244 (April 1981): 41–47.

Grygiel, Fred S., and Anthony J. Zarillo. "Three Mile Island: The New Jersey Regulatory Response [Problems Confronting the N.J. Board of Public Utilities in Its Regulation of Jersey Central Power and Light Company, 25% Owner of the Damaged Nuclear Power Plant]." *Public Utilities Fortnightly* 106 (December 18, 1980): 23–28.

Hartke, Victoria Riess. "The International Fallout from Chernobyl." *Dickinson Journal of International Law* 5 (Spring 1987): 319–343.

Hu, Teh-Wei, and Kenneth S. Slaysman. "Health Related Economic Costs of the Three Mile Island Accident [in Terms of Mental and Physical Adjustments Due to the Breakdown of a Nuclear Power Station Near Harrisburg, PA, March 29, 1979]." *Socio-Economic Planning Sciences* 18:3 (1984): 183–193.

Lange, R., and others. "Dose Estimates from the Chernobyl Accident." *Nuclear Technology* 82 (September 1988): 311–323.

Lanouette, William J. "Nuclear Power—an Uncertain Future Grows Dimmer Still: Three Mile Island [PA] May End Up Revealing Not Only the Industry's Weaknesses But Also Its Persistence and Necessity." *National Journal* 11 (April 28, 1979): 676–686.

______. "One Year Later, the Dangers Persist at Three Mile Island Nuclear Plant [Near Harrisburg, PA]: Owners of the Utility Warn That the Nuclear Regulatory Commission's Cautious Policy Could, Ironically, Increase the Prospect of Further Accidents." *National Journal* 12 (May 3, 1980): 730–732.

Perham, Christine. "EPA's Role at Three Mile Island [Role of the U.S. Environmental Protection Agency Since the Accident at the Nuclear Plant Near Harrisburg, PA, March 28, 1979]." *EPA Journal* 6 (October 1980): 18–20+.

Plattner, Andy. "Congressional Help Urged for Three Mile Island Plant Nuclear Accident Cleanup." *Congressional Quarterly Weekly Report* 39 (March 7, 1981): 403–405.

Staenberg, Marc R. "Financial and Legal Implications of the Three Mile Island Accident." *Nuclear Law Bulletin* (December 1979): 65–71.

Tawil, Jack J., and Dennis L. Strenge. "Using Cost/Risk Procedures to Establish Recovery Criteria Following a Nuclear Reactor Accident." *Health Physics* 52 (February 1987): 157–169.

Walsh, E. J. "Resource Mobilization and Citizen Protest Around Three Mile Island." *Social Problems* 29 (1981): 1–21.

Wheeler, D. A. "Atmospheric Dispersal and Deposition of Radioactive Material from Chernobyl." *Atmospheric Environment* 22:9 (1988): 853–863.

———. "Radioactivity in Rainwater Following the Chernobyl Accident." *Environmentalist* 7 (Spring 1987): 31–34.

Fuel Cycle

Chawla, R., and others. "Effects of Fuel Enrichment on the Physics Characteristics of Plutonium-Fueled Light Water High Converter Reactors." *Nuclear Technology* 73 (June 1986): 296–305.

Cooley, C. R., and J. F. Strahl. "Summary of the United States and the Nuclear Fuel Cycle." *Radioactive Waste Management and the Nuclear Fuel Cycle* 4 (February 1984): 285–305.

Domestic Uranium Mining and Milling Industry: 1983 Viability Assessment. Washington, DC: U.S. Department of Energy, 1984. 191p.

Glackin, James J. "The Dangerous Drift in Uranium Enrichment." *Bulletin of the Atomic Scientists* 32 (February 1976): 22–29.

Hirata, M. "Institutional Arrangements for Nuclear Fuel Cycle Services." *Energy* 9 (September–October 1984): 883–888.

Kocher, D. C. "Performance Objectives for Disposal of Low-Level Radioactive Wastes on the Oak Ridge Reservation." *Radioactive Waste Management and the Nuclear Fuel Cycle* 7 (October 1986): 359–380.

"Management of Inactive Uranium Mill Tailings." *Journal of Environmental Engineering* 112 (June 1986): 490–537.

Papp, R. and K.-D. Closs. "Alternative Fuel Cycle Evaluation in the Federal Republic of Germany." *Nuclear Technology* 72 (March 1986): 312–320.

Papp, R., and H. Loser. "Fuel Reprocessing Versus Direct Disposal of Spent Fuel—A Comparison from the Standpoint of Radiological Safety." *Nuclear Technology* 73 (May 1986): 228–235.

U.S. Congress. House. Committee on Armed Services, Procurement and Military Nuclear Systems Subcommittee. *Management of Commingled Uranium Mill Tailings: Hearings, August 17–18, 1982.* 97th Cong., 2d sess., Washington, DC: GPO, 1982. 673p.

U.S. Congress. House. Committee on Energy and Commerce. Subcommittee on Energy Conservation and Power. *Development of Nuclear Power Fuel Cycles: Report, November 1984.* 99th Cong., 2d sess., Washington, DC: GPO, 1984. 136p.

U.S. Congress. Senate. Committee on Energy and Natural Resources. Subcommittee on Energy Research and Development. *Domestic Uranium Mining Industry and The Department of Energy's Uranium Enrichment Program: Hearings, March 9 and 13, 1987.* 100th Cong., 1st sess., Washington, DC: GPO, 1987. 705p.

U.S. Congress. Senate. Committee on Environment and Public Works. Subcommittee on Nuclear Regulation. *Implementation of the Uranium Mill Tailings Radiation Control Act of 1978: Hearing, June 16, 1981.* 97th Cong., 1st sess., Washington, DC: GPO, 1981. 615p.

U.S. Congress. House. Committee on Interior and Insular Affairs. Subcommittee on Energy and the Environment. *Uranium Mill Tailings Act of 1985; Joint Hearing Before the Subcommittee on Energy and the Environment of the Committee on Interior and Insular Affairs, on H.R. 2236 and S.1004, July 16, 1985.* 99th Cong., 1st sess., Washington, DC: GPO, 1986, 289p.

U.S. Congress. Senate. Committee on Labor and Human Resources. *Radiation Exposure Compensation Act of 1981: Hearing, October 27, 1981, on S.1483, to Amend Title 28 of the United States Code to Make the United States Liable for Damages to Certain Individuals to Certain Uranium Miners, and to Certain Sheep Herds, Due to Certain Nuclear Tests at the Nevada Test Site or Employment in a Uranium Mine, and for Other Purposes.* 97th Cong., 1st sess., Washington, DC: GPO, 1982. 304p.

U.S. Department of Energy. *Annual Status Report on the Uranium Mill Tailings Remedial Action Program.* Washington, DC: 1980.

U.S. Environmental Protection Agency. Office of Radiation Programs. *Regulatory Impact Analysis of Final Environmental Standards for Uranium Mill Tailings at Active Sites.* Washington, DC: GPO, 1983. 5p.

United States Uranium Mining and Milling Industry: A Comprehensive Review. Washington, DC: U.S. Department of Energy, 1984. 103p.

Wilson, Ellen. "Some Left It Hot: More Than 200 Million Tons of Uranium Mill Tailings Lie in Piles Around America: How Will They Be Cleaned Up, Who Will Do It . . . and When?" *Environmental Action* 17 (November-December 1985): 27-32.

Radioactive Waste Disposal

General Works

Alvarez, Robert, and Arjun Makhijani. "Radioactive Waste: Hidden Legacy of the Arms Race: The Military's Vast Quantities of Radioactive Waste Will Threaten Nearby Communities for Centuries and Could Determine the Future of the Nation's Weapons Program." *Technology Review* 91 (August–September 1988): 42–51.

Campbell, John L. "The State and the Nuclear Waste Crisis: An Institutional Analysis of Policy Constraints." *Social Problems (Soc Study Social Problems)* 34 (February 1987): 18–23.

Cohen, Bernard. "The Disposal of Radioactive Wastes from Fission Reactors." *Scientific American* 236 (June 1977): 21–31.

Dornsife, William P. "Classification of Radioactive Materials and Wastes." In Shyamal K. Majumdar and E. Willard Miller, eds., *Management of Radioactive Materials and Wastes: Issues and Progress* (Easton, PA: The Pennsylvania Academy of Science, 1985), 2–9. ISBN 0-9606670-4-0.

Feates, F. S. "UK Policy on Radioactive Wastes." *Radioactive Waste Management and the Nuclear Fuel Cycle* 9:1–3 (1987): 3–17.

Hancock, Don. "The Wasting of America: Target—Nevada, Target—New Mexico." *Workbook (Southwest Research and Info Center)* 13 (January–March): 2–12.

Harrison, J.M. "Disposal of Radioactive Wastes." *Science* 226 (October 5, 1984): 11–14.

"Hazy Radwaste Rules Hold Up Utility Action." *Electrical World* 200 (July 1986): 15–20.

Klingsberg, C., and J. Duguid. "Isolating Radioactive Waste." *American Scientist* 70 (March–April 1982): 182–190.

Marsily, G. de. "Nuclear Waste Disposal: Who Will Make the Final Decision?" *Radioactive Waste Management* 2 (August 1981): 109–121.

Mazur, A., and B. Conant. "Controversy Over a Local Nuclear Waste Repository." *Social Studies of Science* 8 (1978): 235–243.

Merz, Erich R. "Challenges of Waste Handling and Disposal." *Radioactive Waste Management and the Nuclear Fuel Cycle* 7 (April 1986): 107–120.

Monroe, Alison. "Radioactive Wastes: How to Evaluate Solutions." *Workbook (Southwest Research and Info Center)* 6 (March–April 1981): 45–54.

Montange, Charles H. "Federal Nuclear Waste Disposal Policy." *Natural Resources Journal* 27 (Spring 1987): 309–408.

Plattner, Andy. "Nuclear Waste: Concern Growing Over Problem But Congress Can't Agree on Bill." *Congressional Quarterly Weekly Report* 38 (November 1, 1980): 3259–3263+.

New Jersey General Assembly. Environmental Quality Committee. *Public Hearing on Assembly Bill 3019 [and Other Bills] (Legislative and Executive Response with Respect to Storage, Treatment, and Disposal of Radium-Contaminated Soil): Vernon, New Jersey, September 12, 1986.* Trenton, NJ: 1986. 124p.

Nuclear Waste, Monitored Retrievable Storage of Spent Nuclear Fuel: Fact Sheet for Congressional Requestors. Washington, DC: United States General Accounting Office, 1986. 32p.

Nuclear Waste, Quarterly Report on DOE's Nuclear Waste Program as of September 30, 1985. Report to Congressional Requestors. Washington, DC: U.S. General Accounting Office, 1986. 50p.

Containers for Storage

Barkatt, A., and others. "Mechanisms of Defense Waste Glass Dissolution." *Nuclear Technology* 73 (May 1986): 140–164.

Beckman, Petr. "Containing Nuclear Waste: Safe Disposal of Radioactive Waste Is Often Called the Achilles' Heel of Nuclear Power; How Far Along Are We on the Road to Resolving This Problem?" *Reason* 13 (September 1981): 19–26.

Beneson, Robert. "America's Nuclear Waste Backlog." *Editorial Research Reports* (December 4, 1981): 895–912.

Cousens, D. R., and others. "Evaluating Glasses for High-Level Radioactive Waste Immobilization." *Radioactive Waste Management* 2 (December 1981): 143–168.

Flowers, R. H. "Objectives for Radioactive Waste Packaging." *Radioactive Waste Management and the Nuclear Fuel Cycle* 3 (June 1983): 257-277.

Hanson, M. S. "Deposition of Volatile Fission Products in Sintered Metal Filters." *Radioactive Waste Management and the Nuclear Fuel Cycle* 4 (October 1983): 107-127.

Hayward, P. J. "The Use of Glass Ceramics for Immobilizing High-Level Wastes from Nuclear Fuel Recycling." *Glass Technology* 29 (August 1988): 122-136.

———, and others. "Leaching Studies of Sphene Ceramics Containing Substituted Radionuclides." *Nuclear and Chemical Waste Management* 6:1 (1986): 71-80.

Heimann, R. B. "Nuclear Fuel Waste Management and Archaeology: Are Ancient Glasses Indicators of Long Term Durability of Man Made Materials?" *Glass Technology* 27 (June 1986): 96-101.

Implementation Plan for Deployment of Federal Interim Storage Facilities for Commercial Spent Nuclear Fuel. Washington, DC: U.S. Department of Energy, Office of Civilian Radioactive Waste Management, 1988. 15p.

James, L. A. "Cracking of a Nuclear Waste Container Material by Irradiation in a Simulated Groundwater." *Nuclear and Chemical Waste Management* 8:1 (1988): 75-82.

Kamizono, Hiroshi. "Thermal Shock Resistance of High-Level Waste Glass: An Overview." *Radioactive Waste Management and the Nuclear Fuel Cycle* 8:1 (1987): 53-63.

Lewis, R. A., and others. "Analysis for Silicon in Solution in High-Level Waste Glass Durability Studies." *Radioactive Waste Management and the Nuclear Fuel Cycle* 3 (December 1982): 191-198.

Monitored Retrievable Storage Submission to Congress. Washington, DC: U.S. Department of Energy. Office of Civilian Radioactive Waste Management, 1987. 3 vols.

Mosher, Lawrence. "As Nuclear Garbage Heap Grows, So Does The Dispute Over Storing It: The Nuclear Industry and Electric Utilities Say They Need Federal Help to Store Spent Fuel, and Congress and the Energy Department

Are Close to Giving It." *National Journal* 14 (September 11, 1982): 1540-1544.

Murray, R. L. "Radioactive Waste Storage and Disposal." *IEEE Proceedings* 74 (April 1986): 552-579.

"Nuclear Garbage Dumped by the Military: Even with a Stagnating Nuclear Power Industry, We Already Have an Enormous Amount of Nuclear Waste—Thanks to Military Programs." *Business and Society Review* (Fall 1981): 42-48.

O'Brien, John M. "International Auspices for the Storage of Spent Nuclear Fuel as a Nonproliferation Measure." *Natural Resources Journal* 21 (October 1981): 857-894.

Plécăs, I. B., L. J. L. Mihajlović, and A. M. Kostadinović. "Optimization of Concrete Containers Composition in Radioactive Waste Technology." *Radioactive Waste Management and the Nuclear Fuel Cycle* 6 (June 1985): 161-176.

Vecchio, F. J., and J. A. Sato. "Drop, Fire, and Thermal Testing of a Concrete Nuclear Fuel Container." *ACI Structural Journal* 85 (July-August 1988): 374-383.

Geologic Considerations

Arthur, W. J., III, and O. D. Markham. "Small Mammal Burrowing as a Radionuclide Transport Vector at a Radioactive Waste Disposal Area in Southeastern Utah." *Journal of Environmental Quality* 12 (1983): 117-122.

______, and others. "Radiation Dose to Small Mammals Inhabiting a Solid Radioactive Waste Disposal Area." *Journal of Applied Ecology* 23 (1986): 13-26.

Baeyens, B., A. Maes, and A. Cremers. "Aging Effects in Boom Clay." *Radioactive Waste Management and the Nuclear Fuel Cycle* 6 (December 1985): 409-423.

______. "In-Situ Physico-Chemical Characterization of Boom Clay." *Radioactive Waste Management and the Nuclear Fuel Cycle* 6 (December 1985): 391-408.

Bonne, A., and R. Heremans. "A Decade of Research and Development Studies on the Disposal of the High Level and Alpha-Bearing Waste in a Deep Clay Formation." *Radioactive Waste Management and the Nuclear Fuel Cycle* 6 (December 1985): 277-291.

Bourke, P. J., and P. C. Robinson. "Comparison of Thermally Induced and Naturally Occurring Waterborne Leakages from Hard Rock Depositories for Radioactive Waste." *Radioactive Waste Management* 1 (May 1981): 365–380.

Brookins, D. G. "Geochemistry of the Dakota Formation of Northwestern New Mexico: Relevance to Radioactive Waste Studies." *Nuclear Technology* 59 (December 1982): 420–428.

Burgess, A., and others. "Geological Engineering Aspects of the Conceptual Design of a Radioactive Waste Vault in Hard Crystalline Rock." *CIM Bulletin* 73 (July 1980): 62–72.

Cameron, Francis X. "Human Intrusion into Geologic Repositories for High-Level Radioactive Waste: Potential and Prevention." *Radioactive Waste Management* 2 (December 1981): 179–187.

Carlsen, Lars, Walther Batsberg and Bror Skytte Jensen. "Chalk Formations as Natural Barriers Towards Radionuclide Migration." *Radioactive Waste Management and the Nuclear Fuel Cycle* 6 (June 1985): 121–130.

Chapman, Neil, and Ferruccio Gera. "Disposal of Radioactive Wastes in Italian Clays: Mined Depository or Deep Boreholes?" *Radioactive Waste Management and the Nuclear Fuel Cycle* 6 (March 1985): 51–78.

Chiantore, Vittorio, and Ferruccio Gera. "Fracture Permeability of Clays: A Review." *Radioactive Waste Management and the Nuclear Fuel Cycle* 7 (August 1986): 253–277.

Cohen, B. L. "Critique of the National Academy of Sciences Study of the Isolation System for Geologic Disposal of Radioactive Waste." *Nuclear Technology* 70 (September 1985): 433–440.

Crowe, Bruce M., Mark E. Johnson, and Richard J. Beckman. "Calculation of the Probability of Volcanic Disruption of a High-Level Radioactive Waste Repository Within Southern Nevada, USA." *Radioactive Waste Management and the Nuclear Fuel Cycle* 3 (December 1982): 167–190.

D'Alessandro, Marco, and Ferruccio Gera. "Geological Isolation of Radioactive Waste in Clay Formations: Fractures and Faults as Possible Pathways for Radionuclide Migration." *Radioactive Waste Management and the Nuclear Fuel Cycle* 7 (October 1986): 381–406.

Daniel, D. E. "Shallow Land Burial of Low-Level Radioactive Waste." *Journal of Geotechnical Engineering* 109 (December 1983): 1641–1646.

Day, D. A. "Deep Storage of Nuclear Waste: Structural Issues." *Proceedings of the American Society of Civil Engineers* 106 (EY2 no. 15746) (October 1980): 201–212.

DeMarsily, G., and others. "Nuclear Waste Disposal: Can the Geologist Guarantee Isolation?" *Science* 197 (August 5, 1977): 519–527.

Groves, C. R., and B.L. Keller. "Ecological Characteristics of Small Mammals on a Radioactive Waste Disposal Area in Southeastern Idaho." *American Midland Naturalist* 109 (1983): 253–265.

Henrion, P. N., and others. "Migration of Radionuclides in Boom Clay." *Radioactive Waste Management and the Nuclear Fuel Cycle* 6 (December 1985): 313–359.

Irish, Everrett R., and Carl R. Cooley. "Status of Technologies Related to the Isolation of Radioactive Wastes in Geologic Repositories." *Radioactive Waste Management* 1 (September 1980): 121–146.

Levy, P. W. "Radiation Damage Studies on Natural Rock Salt from Various Geological Localities of Interest to the Radioactive Waste Disposal Program." *Nuclear Technology* 60 (February 1983): 231–243.

Li, W. T., and C. L. Wu. "Analysis of Creep for Nuclear Waste Storage in a Salt Formation." *Nuclear Technology* 61 (May 1983): 344–352.

McKinley, Ian G., and Julia M. West. "Radionuclide Sorption on Argillaceous Rocks from Harwell, Oxfordshire." *Radioactive Waste Management and the Nuclear Fuel Cycle* 4 (February 1984): 379–399.

Neerdael, B., and P. Manfroy. "Geotechnical Characterization of Clay at Great Depth and Its Connection to Tunnel Construction." *Radioactive Waste Management and the Nuclear Fuel Cycle* 6 (December 1985): 293–312.

Pigford, T. H. "Geological Disposal of Radioactive Waste." *Chemical Engineering Progress* 78 (March 1982): 18–26.

Reynolds, Timothy D., and John W. Laundré. "Vertical Distribution of Soil Removed by Four Species of Burrowing Rodents in Disturbed and Undisturbed Soils." *Health Physics* 54 (April 1988): 445–450.

Saltelli, A., and F. Antonioli. "Radioactive Waste Disposal in Clay Formations: A Systematic Approach to the Problem of Fractures and Faults Permeability." *Radioactive Waste Management and the Nuclear Fuel Cycle* 6 (June 1985): 101–120.

Seitz, M. G., and others. "Migratory Properties of Some Nuclear Waste Elements in Geologic Media." *Nuclear Technology* 44 (July 1979): 284–296.

Silva, A. J., and others. "Geotechnical Properties of Sediments from the North Central Pacific and Northern Bermuda Rise." *Marine Geotechnology* 5:3–4 (1984): 235–256.

Verkerk, B. "Comparison of Long-Term Release Consequences for Spent Fuel and Vitrified Waste Repositories in Salt Formation." *Radioactive Waste Management* 1 (May 1981): 337–357.

West, Julia M., Nicholas Christofi, and Ian G. McKinley. "An Overview of Recent Microbiological Research Relevant to the Geological Disposal of Nuclear Waste." *Radioactive Waste Management and the Nuclear Fuel Cycle* 6 (March 1985): 79–95.

Witherspoon, P. A., and others. "Geologic Storage of Radioactive Waste: Field Studies in Sweden." *Science* 211 (February 27, 1981): 894–900.

Wood, B. J. "Backfill Performance Requirements: Estimates from Transport Models." *Nuclear Technology* 59 (December 1982): 390–404.

Wood, M. I., and W. E. Coons. "Basalt as a Potential Waste Package Backfill Component in a Repository Located Within the Columbia River Basalt." *Nuclear Technology* 59 (December 1982): 409–419.

High-Level Waste Disposal

Come, B. "Performances of High-Level Radioactive Waste Forms with a View to Geological Disposal." *Radioactive Waste Management and the Nuclear Fuel Cycle* 7 (March 1986): 83–95.

Gray, W. J. "Volatility of Some Potential High-Level Radioactive Waste Forms." *Radioactive Waste Management* 1 (September 1980): 147–169.

Grover, J. R. "High-Level Waste Solidification." *Radioactive Waste Management* 1 (May 1980): 2–12.

Hambley, D. F., and J. R. Morris. "Designing Shafts for Handling High-Level Radioactive Wastes in Mined Geologic Repositories." *Nuclear Technology* 80 (March 1988): 476–482.

Kocher, D. C., and others. "A Perspective on Demonstrating Compliance with Standards for Disposal of High-Level Radioactive Wastes." *Radioactive Waste Management and the Nuclear Fuel Cycle* 6 (March 1985): 1–18.

"Monitored Retrievable Storage (MRS) Facility for High-Level Radioactive Waste." *Forum for Applied Research and Public Policy* 1 (Spring 1986): 10–54.

Prij, J. "On the Time Dependent Behavior of a Cylindrical Salt Dome with a High-Level Waste Repository." *Nuclear Technology* 80 (March 1988): 462–475.

Ross, B. "Scenarios in Performance Assessment of High-Level Waste Repositories." *Radioactive Waste Management and the Nuclear Fuel Cycle* 7 (March 1986): 47–61.

Schilling, A. Henry. "High-Level Radioactive Waste: Washington [State's] Prospects." *Washington Public Policy Notes* 11 (Spring 1983): 1–6.

Sousselier, Y. "Criteria for High-Level Waste Disposal." *Radioactive Waste Management* 1 (May 1981): 359–364.

U.S. Environmental Protection Agency. Office of Radiation Programs. *Environmental Standards for Management and Disposal of Spent Nuclear Fuel, High-Level Transuranic Radioactive Wastes: Draft Regulatory Impact Analysis for 40 CFR 191 [Title 40, Code of Federal Regulations, Part 191].* Washington, DC: GPO, 1982. 84p.

U.S. Congress. House. Committee on Interior and Insular Affairs. Subcommittee on Energy and the Environment. *High-Level Nuclear Waste Management: Oversight Hearing, June 17, 1982.* 97th Cong., 2d sess., Washington, DC: GPO, 1982. 148p.

U.S. Congress. Office of Technology Assessment. *Managing Commercial High-Level Radioactive Waste: Summary.* Washington, DC: GPO, 1982. 64p.

U.S. Department of Commerce. National Bureau of Standards. *An Analysis of the Requirements for a Computer Assisted Database for Reviews and Evaluations on High Level Waste Data.* NBSIR, 86-3363. Gaithersburg, MD: 1986. 57p.

Low-Level Waste Disposal

Breiner, E. M. "Low Level Radioactive Waste Disposal—Technology and Public Policy." *Journal of Environmental Sciences* 29 (July/August 1986): 47–56.

Brenneman, Faith N. "A Review of Low-Level Radioactive Waste Compacts on a National Level." In Shyamal K. Majumdar and E. Willard Miller, eds., *Management of Radioactive Materials and Wastes: Issues and Progress* (Easton, PA: The Pennsylvania Academy of Science, 1985), 29–41. ISBN 0-9606670-4-0.

Brill, David R. "Disposal Problems of Low-Level Radioactive Wastes and Implications for Hospitals." In Shyamal K. Majumdar and E. Willard Miller, eds., *Management of Radioactive Materials and Wastes: Issues and Progress* (Easton, PA: The Pennsylvania Academy of Science, 1985), 98–106. ISBN 0-9606670-4-0.

Choi, Yearn Hong. "Issues of New Federalism in Low-Level Radioactive Waste Management: Cooperation or Confusion?" *State Government* 57:1 (1984): 13–20.

———. "Low-Level Radioactive Waste Management: Federal-State Cooperation or Confusion?" *Journal of Environmental Sciences* 27 (July–August 1984): 41–46.

Cohen, B. L. "A Generic Probabilistic Risk Assessment for Low-Level Waste Burial Grounds." *Nuclear and Chemical Waste Management* 5:1 (1984): 39–47.

Coleman, Joseph A. "Low-Level Radioactive Waste Management Technology Development." In Shyamal K. Majumdar and E. Willard Miller, eds., *Management of Radioactive Materials and Wastes: Issues and Progress* (Easton, PA: The Pennsylvania Academy of Science, 1985), 10–14. ISBN 0-9606670-4-0.

DiSibio, Ralph. "Operation of a Low-Level Waste Disposal Facility and How To Prevent Problems in Future Facilities." In Shyamal K. Majumdar and E. Willard Miller, eds., *Management of Radioactive Materials and Wastes: Issues and Progress* (Easton, PA: The Pennsylvania Academy of Science, 1985), 137–141. ISBN 0-9606670-4-0.

Jarrett, A. R. "Effective Water Management for Low-Level Radioactive Waste Repositories." In Shyamal K. Majumdar and E. Willard Miller, eds., *Management of Radioactive Materials and Wastes: Issues and Progress* (Easton, PA: The Pennsylvania Academy of Science, 1985), 15–28. ISBN 0-9606670-4-0.

Jones, Woodrow, and Pamela M. Berger. "Low-Level Radioactive Waste Disposal Policies and State Governments." In Dennis R. Judd, ed., *Public Policy Across States and Communities.* Greenwich, CT: JAI Press, 1985, 109–126.

Köster, R., and W. Bechthold. "Conditioning of Low and Intermediate Level Wastes." *Radioactive Waste Management and the Nuclear Fuel Cycle* 7 (April 1986): 151–164.

Laser, M. "Volume Reduction of Low Level Solid Radioactive Waste by Incineration and Compaction in the Federal Republic of Germany." *Radioactive Waste Management and the Nuclear Fuel Cycle* 7 (April 1986): 165–180.

Moore, T. "The Great State of Uncertainty in Low-Level Waste Disposal." *EPRI Journal* 10 (March 1985): 22–29.

______. "Remote Scanning of Low-Level Waste." *EPRI Journal* 11 (June 1986): 22–27.

Prochaska, J. R. "Low-Level Radioactive Waste Disposal Compacts." *Virginia Journal of Natural Resources Law* 5 (1986): 383–411.

Pushchak, Ronald, and Ian Burton. "Risk and Prior Compensation in Siting Low-Level Nuclear Waste Facilities: Dealing with the NIMBY [Not in My Back Yard] Syndrome." *Plan Canada* 23 (December 1983): 68–79.

Stubbs, Anne D. "The Northeast Low-Level Waste Compact: Regional Cooperation in Low-Level Waste Management." In Shyamal K. Majumdar and E. Willard Miller, eds., *Management of Radioactive Materials and Wastes: Issues and Progress* (Easton, PA: The Pennsylvania Academy of Science, 1985), 42–53. ISBN 0-9606670-4-0.

U.S. Congress. *Low-Level Radioactive Waste Disposal: Joint Hearing, October 8, 1985, Before the Subcommittee on Energy Research and Development of the Committee on Energy and Natural Resources and the Subcommittee on Nuclear Regulation of the Committee on Environment and Public Works on S.1517 and S.1578.* 99th Cong. 1st sess., Washington, DC: GPO, 1986. 681p.

Yalow, Rosalyn S. "Disposal of Low-Level Radioactive Biomedical Wastes: A Problem in Regulation, Not Science." *Radioactive Waste Management* 1 (May 1986): 319–323.

Management

Abrams, N. E., and J. R. Primack. "Helping the Public Decide: The Case of Radioactive Waste Management." *Environment* 22 (1980): 14–20.

Bailey, G. "Radioactive Waste Management at Dounreay." *Radioactive Waste Management and the Nuclear Fuel Cycle* 9:1–3 (1987): 27–50.

Carter, T. J. "Radioactive Waste Management Practices at a Large Canadian Electrical Utility." *Radioactive Waste Management* 2 (June 1982): 381–412.

Cummings, R. G., ed. "Symposium on the Management of Nuclear Wastes." *Natural Resources Journal* 21 (October 1981): 693–894.

Derrington, J. A. "Toxic and Radioactive Waste Management." *Journal of Professional Issues in Engineering* 114 (October 1988): 463–481.

Detilleux, E. "Radioactive Waste Management in Belgium." *Radioactive Waste Management and the Nuclear Fuel Cycle* 4 (February 1984): 253–263.

Feates, F. S. "U.K. Policy on Radioactive Wastes." *Radioactive Waste Management and the Nuclear Cycle* 9:1–3 (1987): 3–17.

Fellingham, L. R. "Radioactive Waste Management R & D at Harwell." *Radioactive Waste Management and the Nuclear Fuel Cycle* 9:1–3 (1987): 151–181.

Ferrero, J. L., and others. "Atmospheric Radioactivity in Valencia, Spain, Due to the Chernobyl Reactor Accident." *Health Physics* 53 (November 1987): 519–524.

Finn, Daniel P. "Nuclear Waste Management Activities in the Pacific Basin and Regional Cooperation on the Nuclear Fuel Cycle." *Ocean Development and International Law* 13:2 (1983): 213–246.

Flax, Susan Jo. "Radioactive Waste Management." *Harvard Environmental Law Review* 5:2 (1981): 259–295.

Forrest, C. "Radioactive Waste Management." *Nuclear Energy* 24 (April 1987): 79–83.

Friedman, Robert S. "Political Considerations of Nuclear Waste Disposal Policy." In Shyamal K. Majumdar and E. Willard Miller, eds., *Management of Radioactive Materials and Wastes: Issues and Progress* (Easton, PA: The Pennsylvania Academy of Science, 1985), 203–215. ISBN 0-9606670-4-0.

Gandellini, A. "The Management and Disposal of Radioactive Wastes in Italy." *Radioactive Waste Management and the Nuclear Fuel Cycle* 4 (February 1984): 265–268.

Grossman, P. Z., and E. S. Cassedy. "Cost-Benefit Analysis of Nuclear Waste Disposal: Accounting for Safeguards." *Science, Technology, and Human Values* 10 (Fall 1985): 47–54.

Grover J. R. "The Management and Disposal of Low and Intermediate-Level Radioactive Wastes in the U.K." *Radioactive Waste Management and the Nuclear Fuel Cycle* 4 (February 1984): 269–283.

Haigh, C. P., and J. A. Luke. "Recent Developments in Power Station Waste Management Procedures." *Radioactive Waste Management and the Nuclear Fuel Cycle* 9:1–3 (1987): 71–84.

Hill, Sir John. "After the Great Nuclear Debate [in Terms of Public Perception of Dangers Involved and Nuclear Waste Management]." *Atom (Gt. Brit.)* (January 1981): 2–8.

Kearney, Richard C., and Robert B. Garey. "American Federalism and the Management of Radioactive Wastes." *Public Administration Review* 42 (January–February 1982): 14–24.

Kesson, S. E., and A. E. Ringwood. "Safe Disposal of Spent Nuclear Fuel." *Radioactive Waste Management and the Nuclear Fuel Cycle* 4 (October 1983): 159–174.

King, Joseph C. "MRS and the Local Community: The Oak Ridge Area Response." *Forum for Applied Research and Public Policy* 1 (Spring 1986): 42–49.

Kittel, J. H. "Nuclear Waste Management—Issues and Progress." *Journal of Environmental Sciences* 27 (March–April 1984) 34–41.

Lavie, J. M., and M. N. Moscardo. "Radioactive Waste Management in France." *Radioactive Waste Management and the Nuclear Fuel Cycle* 4 (February 1984): 235–251.

Lee, K. N. "Federalist Strategy for Nuclear Waste Management." *Science* 208 (May 16, 1980): 679–684.

Lehtinen, R., P. Silvennoinen, and J. Vira. "Allowing for Uncertainties in the Expected Waste Management Costs." *Radioactive Waste Management* 1 (May 1980): 43–55.

Mission Plan for the Civilian Radioactive Waste Management Program. Washington, DC: U.S. Department of Energy, Office of Civilian Radioactive Waste Management, 1985. Vol. 1. 457p.

Murauskas, G. Thomas, and Fred M. Skelley. "Local Political Responses to Nuclear Waste Disposal." *Cities* 3 (May 1986): 157–162.

Nilsson, Lars B., and Claes Thegerström. "The Swedish Nuclear Waste Management Program." *Radioactive Waste Management and the Nuclear Fuel Cycle* 4 (February 1984): 221–234.

Olds, F. C. "Nuclear Waste Management." *Power Engineering* 89 (November 1985): 30–36.

______. "Nuclear Waste Management." *Power Engineering* 85 (May 1981): 48–56.

Public Comments on the Draft Mission Plan for the Civilian Waste Management Program. Washington, DC: U.S. Department of Energy, Office of Radioactive Waste Management, 1985. Vol. 3. 719p.

Raudenbush, M. H. "Looking at Waste Management Worldwide." *Nuclear Engineering International* 28 (August 1983): 30–33.

Record of Responses to Public Comments on the Draft Mission Plan for the Civilian Radioactive Waste Management Program. Washington, DC: U.S. Department of Energy, Office of Civilian Waste Management, 1985. 354p.

Richter, D., and S. Fareeduddin. "Developing Guidelines in Managing Radioactive Waste." *IAEA Bulletin* 24 (June 1982): 6–12.

Rometsch, R. "Cooperative Management of Nuclear Waste Disposal in Switzerland." *Radioactive Waste Management and the Nuclear Fuel Cycle* 4 (February 1984): 213–220.

Roy, Rustum. "The Technology of Nuclear Waste Management." *Technology Review* 83 (April 1981): 39–43+.

U.S. Congress. House. Committee on Energy and Commerce. Subcommittee on Energy Conservation and Power. *Nuclear Waste Management: Hearing, June 16, 1981.* 97th Cong., 1st sess., Washington, DC: GPO, 1981. 61p.

U.S. Congress. Office of Technology Assessment. "Managing Commercial High-Level Radioactive Waste." *Radioactive Waste Management and the Nuclear Fuel Cycle* 3 (June 1983): 279–345.

U.S. Congress. Senate. Committee on Governmental Affairs. Subcommittee on Energy, Nuclear Proliferation and Federal Services. *Nuclear Waste Management Reorganization Act of 1979: Hearings, July 5, 1979–February 13, 1980, and S.742 to Effect Certain Reorganization of the Federal Government to Strengthen Federal Programs and Policies with Respect to Nuclear Waste Management.* 96th Cong., 1st sess., Washington, DC: GPO, 1980. 247p.

U.S. General Accounting Office. *Nuclear Waste: Unresolved Issues Concerning Hanford's Waste Management Practices: Report to Congressional Requestors.* Washington, DC: 1987. 69p.

Yasinsky, John B., and Charles R. Bolmgren. "Radioactive Waste Management—A Manageable Task." In Shyamal K. Majumdar and E. Willard Miller, eds., *Management of Radioactive Materials and Wastes: Issues and Progress* (Easton, PA: The Pennsylvania Academy of Science, 1985), 73–96. ISBN 0-9606670-4-0.

Oceanic Disposal

Baxter, M. S., and S. R. Aston. "Iodine-129 in High-Activity Nuclear Wastes: An Assessment of the Deep-Ocean Disposal Option." *Radioactive Waste Management and the Nuclear Fuel Cycle* 3 (September 1982): 47–55.

______, B. Economides, and S. R. Aston. "Deep-Ocean Disposal of High-Activity Nuclear Wastes: A Conservative Assessment of the Seafood Critical Pathway." *Radioactive Waste Management and the Nuclear Fuel Cycle* 5 (June 1984): 39–52.

Bewers, J. M., and C. J. R. Garrett. "Analysis of the Issues Related to Sea Dumping of Radioactive Wastes." *Marine Policy* 11 (April 1987): 105–124.

Boehmer-Christiansen, Sonja. "An End to Radioactive Waste Disposal 'at Sea'." *Marine Policy* 10 (April 1986): 119–131.

Camplin, W. C., and M. D. Hill. "Sea Dumping of Solid Radioactive Waste: A New Assessment." *Radioactive Waste Management and the Nuclear Fuel Cycle* 7 (August 1986): 233–251.

Coates, R. F. W. "A Deep-Ocean Penetrator Telemetry System." *IEEE Journal of Oceanic Engineering* 13 (August 1988): 55–63.

Cohen, B. "Ocean-Dumping of High-Level Waste: An Acceptable Solution We Can Guarantee." *Nuclear Technology* 47 (January 1980): 163–172.

DeMarsily, G., and others. "Application of Systems Analysis to the Disposal of High Level Waste in Deep Ocean Sediments." *Radioactive Waste Management and the Nuclear Fuel Cycle* 3 (December 1982): 199–213.

Guarascio, John A. "The Regulation of Ocean Dumping After City of New York vs. Environmental Protection Agency." *Boston College Environmental Affairs Law Review* 12 (Summer 1985): 701–741.

Hickox, C. E., and others. "Analysis of Heat Mass Transfer in Subseabed Disposal of Nuclear Waste." *Marine Geotechnology* 5:3–4 (1984): 335–360.

Higgo, J. J. W., L. V. C. Rees, and D. S. Cronan. "Sorption of Plutonium by Deep-Sea Sediments." *Radioactive Waste Management and the Nuclear Fuel Cycle* 6 (June 1985): 143–159.

Holcomb, W. F., and others. "USEPA Radioactive Waste Disposal Standards: Issued and Under Development." *Nuclear and Chemical Waste Management* 8:1 (1988): 3–12.

Hollister, C. D., and others. "Subseabed Disposal of Nuclear Wastes." *Science* 213 (September 18, 1981): 1321–1363.

Hunsaker, Carolyn T. "Ocean Dumping of Low-Level Radioactive Waste: Review of U.S. Laws and International Agreements." *Environmental Forum* 3 (November 1984): 24–31.

Kelly, J. E., and C. E. Shea. "Subseabed Disposal Program for High-Level Radioactive Waste; Public Response." *Oceanus* 25 (Summer 1982): 42–53.

Laine, E. P., and others. "Evaluation Stability and Predictability of Sediment of the Northern Bermuda Rise by the Subseabed Disposal Program." *Marine Geotechnology* 5:3 (1984): 215–233.

Lewis, J. B. "The Case for Deep-Sea Disposal of Low-Level Solid Radioactive Wastes." *Nuclear Energy* 22 (February 1983): 47–51.

Miles, M. E., and others. "Environmental Monitoring and Disposal of Radioactive Waste from Naval Nuclear Vessels and Support Facilities in 1978." *Nuclear Safety* 20 (July 1979): 446–458.

Murray, C. N. "Marine Radiological Studies Related to Seabed Disposal." *Radioactive Waste Management* 2 (August 1981): 83–99.

———, and D. A. Stanners. "Development of an Assessment Methodology for the Disposal of High-Level Radioactive Waste into Deep Ocean Sediments." *Radioactive Waste Management* 2 (March 1982): 239–293.

Needler, G. T., and W. L. Templeton. "Radioactive Waste: The Need to Calculate an Oceanic Capacity." *Oceanus* 24 (Spring 1981): 60–67.

Park, P. Kitho. "Disposal of Radioactive Wastes in the Ocean." *Sea Technology* 25 (January 1984): 62–67.

Reda, R. J., and others. "Intergranular Attack Observed in Radiation-Enhanced Corrosion of Mild Steel." *Corrosion* 44 (September 1988): 632–637.

Searle, R. C. "Guidelines for the Selection of Sites That Might Prove Suitable for Radioactive Waste Disposal On Or Beneath the Ocean Floor." *Nuclear Technology* 64 (February 1984): 166–174.

Solomon, K. A. "Sources of Radioactivity in the Ocean Environment from Low Level Waste to Nuclear Powered Submarines." *Journal of Hazardous Materials* 18 (June 1988): 255–262.

———, and M. Triplett. "A Survey of Monitoring Technology for Subseabed Disposal of Radioactive Waste." *Journal of Hazardous Materials* 10 (July 1985): 205–226.

Templeton, W. L. "Dumping Packaged Low Level Wastes in the Deep Ocean." *Nuclear Engineering International* 27 (February 1982): 36–41.

———, and A. Preston. "Ocean Disposal of Radioactive Wastes." *Radioactive Waste Management and the Nuclear Fuel Cycle* 3 (September 1982): 75–113.

Van Dyke, Jon M. "Ocean Disposal of Nuclear Wastes." *Marine Policy* 12 (April 1988): 82–95.

Welsch, Hubertus. "The London Dumping Convention and Subseabed Disposal of Radioactive Waste." In Jast Delbrück and others, eds., *German Yearbook of International Law, 1985*. Berlin, Germany (Federal Republic): Duncker und Humblot, 1985, pp. 332–354.

Safety

Ahlstrom, P. E., and others. "Safe Handling and Storage of High Level Radioactive Waste." *Radioactive Waste Management* 1 (May 1980): 57–103.

Bertozzi, G., and M. D'Alessandro. "A Probabilistic Approach to the Assessment of the Long-Term Risk Linked to the Disposal of Radioactive Waste in Geological Repositories." *Radioactive Waste Management and the Nuclear Fuel Cycle* 3 (December 1982): 117-136.

Campbell, John L. "The State and the Nuclear Waste Crisis: An Institutional Analysis of Policy Constraints." *Social Problems (Soc. Study Social Problems)* 34 (February 1987): 18-33.

Chang, Soon Heung, and Won Jin Cho. "Risk Analysis of Radioactive Waste Repository Based on the Time Dependent Hazard Rate." *Radioactive Waste Management and the Nuclear Fuel Cycle* 5 (June 1985): 63-80.

Courtney, J. C., and others. "Radiological Safety Analysis of the Hot Fuel Examination Facility/South." *Nuclear Technology* 73 (April 1986): 30-41.

D'Alessandro, M., and others. "Probability Analysis of Geological Processes: A Useful Tool from the Safety Assessment of Radioactive Waste Disposal." *Radioactive Waste Management* 1 (May 1980): 25-42.

Davis, Joseph A. "Nuclear Waste: An Issue That Won't Stay Buried." *Congressional Quarterly Weekly Report* 45 (March 14, 1987): 451-456.

Downey, Gary L. "Federalism and Nuclear Waste Disposal: The Struggle Over Shared Decision Making." *Journal of Policy Analysis and Management* 5 (Fall 1985): 73-99.

Eames, R. P. "Radiation Protection: Some of the Problems of Radiation and the Instruments and Equipment Available for Protecting Personnel." *Safety Engineering* (December 1949): 33-38.

Ehrlich, Robert, and James Ring. "Fallout Sheltering: Is It Feasible?" *Health Physics* 52 (March 1987): 267-280.

Faltermayer, Edmund. "Taking Fear Out Of Nuclear Power." *Fortune* 118 (August 1, 1988): 105-118.

Ginniff, M. E. "Implementation of UK Policy and Strategy on Radioactive Waste Disposal." *Nuclear Energy* 24 (April 1985): 99-104.

Marchetti, Stephen. "Management of Radioactive Spill Prevention." In Shyamal K. Majumdar and E. Willard Miller, eds., *Management of*

Radioactive Materials and Wastes: Issues and Progress (Easton, PA: The Pennsylvania Academy of Science, 1985), 226–236. ISBN 0-9606670-4-0.

Memmert, G., and others. "The German Project-Safety Studies for Nuclear Waste Management: Development of a Safety Assessment Methodology for Final Disposal of Nuclear Waste in a Salt Dome." *Radioactive Waste Management and the Nuclear Fuel Cycle* 7 (April 1986): 209–224.

Newman, M. M. "Nuclear Waste Office Pushes Toward 1998 Goal." *EPRI Journal* 9 (December 1984): 39–44.

"The Nuclear Legacy—How Safe Is It? [Emphasis on the Nuclear Waste Policy Act of 1982 and Its Implementation by the Department of Energy]." *Workbook (Southwest Research and Info Center)* 8 (July–October 1983): 149–172.

Piet, S. J. "Approaches to Achieving Inherently Safe Fusion Power Plants." *Fusion Technology* 10 (July 1986): 7–30.

Pochin, Sir Edward E. "The Evolution of Radiation Protection Criteria." *Nuclear Energy* 25 (February 1986): 19–27.

Roll, David F. "Toxic Agent and Radiation Control: Progress Toward Objectives for the Nation for the Year 1990." *Public Health Reports* 103 (July–August 1988): 342–347.

Ross, Benjamin. "Criteria for Long-Term Safety of Radioactive Wastes: A Proposal." *Radioactive Waste Management and the Nuclear Fuel Cycle* 4 (October 1983): 175–193.

Ryan, Michael T., and David G. Ebenhack. "The Management of Radioactive Materials Spills." In Shyamal K. Majumdar and E. Willard Miller, eds., *Management of Radioactive Materials and Wastes: Issues and Progress* (Easton, PA: The Pennsylvania Academy of Science, 1985), 243–250. ISBN 0-9606670-4-0.

Shuey, Chris, and others. "The 'Costs' of Uranium: Who's Paying with Lives, Lands, and Dollars." *Workbook (Southwest Research Info Center)* 10 (July/September 1985): 102–117.

Thompson, B. G. J. "The Development of Procedures for the Risk Assessment of Underground Disposal of Radioactive Wastes: Research Funded by the Department of Environment 1982-1987." *Radioactive Waste Management and the Nuclear Fuel Cycle* 9:1-3 (1987): 215-256.

U.S. Congress. House. Committee on Armed Services. Procurement and Military Nuclear Systems Subcommittee. *EPA Radon and Radionuclides Emission Standards: Hearing, October 6, 1983.* 98th Cong., 1st sess., Washington, DC: GPO, 1983. 652p.

U.S. Congress. House. Committee on Foreign Affairs. *The International Atomic Energy Agency (IAEA): Improving Safeguards: Hearings, March 3 and 18, 1982, Before the Subcommittees on International Security and Scientific Affairs and on International Economic Policy and Trade.* 97th Cong., 2d sess., Washington, DC: GPO, 1982. 287p.

______. *International Physical Security Standards for Nuclear Materials Outside the United States: Reports.* 100th Cong., 2d sess., Washington, DC: GPO, 1988. 246p.

U.S. Congress. House. Committee on Interior and Insular Affairs. Subcommittee on Oversight and Investigations. *Emergency Preparedness for Radiological Accidents: The Issue of Potassium Iodide: Oversight Hearing, March 5, 1982.* 97th Cong., 2d sess., Washington, DC: GPO, 1983. 391p.

U.S. Congress. House. Committee on Science and Technology. Subcommittee on Energy Research and Production. *The Technology for an Ultrasafe Reactor: Hearing, September 17, 1986.* 99th Cong., 2d sess., Washington, DC: GPO, 1987. 205p.

U.S. Congress. Senate. Committee on Governmental Affairs. Subcommittee on Energy, Nuclear Proliferation, and Federal Services. *DOE's Safety and Health Program for Enrichment Plant Workers: Hearing, July 21, 1980.* 96th Cong., 2d sess., Washington, DC: GPO, 1980. 377p.

U.S. Congress. Senate. Committee on Government Affairs. Subcommittee on Energy, Nuclear Proliferation and Government Processes. *Radiation Protection Management Act of 1982: Hearing, April 29, 1982, on S.2284, to Insure Adequate Protection of Workers, the General Public, and the Environment from Harmful Radiation Exposure, to Establish Mechanisms for Effective Coordination Among the Various Federal Agencies Involved in Radiation Protection Activities, to Develop a Coordinate Radiation Research Program and for Other Purposes.* 97th Cong., 2d sess., Washington, DC: GPO, 1982. 150p.

______. *International Nuclear Safety Concerns: Hearing, May 8, 1986.* 99th Cong., 2d sess., Washington, DC: GPO, 1986. 60p.

U.S. Congress. Senate. Committee on the Judiciary. Subcommittee on Criminal Law. *Implementing the Convention for the Physical Protection of Nuclear Material: Hearing, March 24, 1982, on S.1146.* 97th Cong., 2d sess. Washington, DC: GPO, 1982, 83p.

Webb, G. A. M. "The Tripod on Which the Safety Assessment of Disposal Rests." *Radioactive Waste Management* 1 (September 1980): 113–119.

Weinberg, Alvin M., and John O. Blomeke. "How to Dispose of the Garbage of the Atomic Age." *Across the Board* 19 (September 1982): 26–37 (October 1982): 36–47.

Site Selection

Björnstad, David J., and Ernest Goss. "Issues in the Use of Payments in Lieu of Taxes to Provide Nuclear Waste Facility Siting Incentives." *Radioactive Waste Management* 2 (December 1981): 125–142.

Bord, Richard J. "Problems in Siting Low Level Radioactive Wastes: A Focus on Public Participation." In Shyamal K. Majumdar and E. Willard Miller, eds., *Management of Radioactive Materials and Wastes: Issues and Progress* (Easton, PA: The Pennsylvania Academy of Science), 1985, 189–202. ISBN 0-9606670-4-0.

Bowen, V. T., and C. D. Hollister. "Pre- and Post-Dumping Investigations for Inauguration of New Low-Level Radioactive Waste Dump Sites." *Radioactive Waste Management* 1 (January 1981): 232–269.

Brown, Michael. "The Lower Depths: Underground Injection of Hazardous Wastes." *Amicus Journal* 7 (Winter 1986): 14–23.

Chapman, Neil A., and others. "Site Selection and Characterization for Deep Radioactive Waste Repositories in Britain: Issues and Research Trends into the 1990s." *Radioactive Waste Management and the Nuclear Fuel Cycle* 9:1–3 (1987): 183–213.

Davis, E. C., and others. "Water Diversion at Low-Level Waste Disposal Sites." *Journal of Environmental Engineering* 111 (October 1985): 714–729.

Dayal, R., and others. "Source Term Characterization for the Maxey Flats Low-Level Radioactive Waste Disposal Site." *Nuclear Technology* 72 (February 1986): 158–177.

DiMento, J. F., and others. "Siting Low-Level Radioactive Waste Facilities." *Journal of Environmental Systems* 15 (1985–1986): 19–43.

Doctor, P. G. "The Use of Geostatistics in High Level Radioactive Waste Repository Site Characterization." *Radioactive Waste Management* 1 (September 1980): 193–210.

Gutierrez, Alberto, and Kim Bullerdick. "Underground Storage Tanks and Corrective Action." *Environmental Forum* 4 (September 1985): 33–37.

Holcomb, W. F. "Inventory (1962–1978) and Projections (to 2000) of Shallow Land Burial of Radioactive Wastes at Commercial Sites: An Update." *Nuclear Safety* 21 (May–June 1980): 380–388.

Jacobs, D. G., and R. R. Rose. "Shallow Land Burial of Radioactive Wastes." In Shyamal K. Majumdar and E. Willard Miller, eds., *Management of Radioactive Materials and Wastes: Issues and Progress* (Easton, PA: The Pennsylvania Academy of Science, 1985), 54–72. ISBN 0-9606670-4-0.

Jacobs, Sally. "The Crisis Ahead in Waste Disposal: As the Deadline for Establishing a Site Draws Closer, the Northeast Still Remains Without a Hazardous Waste Plan [Radioactive Waste Disposal Plan Required of All States by 1986 under the Low Level Radioactive Waste Policy Act of 1980]." *New England Business* 6 (September 17, 1984): 86–90+.

Kates, R. W., and B. Braine. "Locus, Equity, and the West Valley Nuclear Wastes." In R. E. Kasperson, ed., *Equity Issues in Radioactive Management.* Cambridge, MA: Oelgeschlager, Gunn & Hain, 1983, 94–117.

Paige, Hilliard W., Daniel S. Lipman, and Janice E. Owens. "Assessment of National Systems for Obtaining Local Acceptance of Waste Management Siting and Routing Activities." *Radioactive Waste Management* 2 (August 1981): 1–48.

Payne, B. A., and others. "The Effects on Property Values of Proximity to a Site Contaminated With Radioactive Waste." *Natural Resources Journal* 27 (Summer 1987): 579–590.

Routson, R. C., and others. "Hanford Site Sorption Studies for the Control of Radioactive Wastes: A Review." *Nuclear Technology* 54 (July 1981): 100–106.

Solomon, B. D., and D. M. Cameron. "Nuclear Waste Repository Siting: An Alternative Approach." *Energy Policy* 13 (1985): 564–580.

———, and Fred M. Shelley. "Siting Patterns of Nuclear Waste Repositories." *Journal of Geography* 87 (March–April 1988): 59–71.

U.S. Congress. House. Committee on Interstate and Foreign Commerce. Subcommittee on Energy and Power. *Nuclear Waste Disposal: Hearing, July 25, 1980, on H.R. 5809 [and Other Bills] to Amend the Atomic Energy Act of 1954, to Authorize States to Enter into Agreements or Compacts with Other States for the Establishment of Regional Disposal Sites for Low-Level Radioactive Waste, to Establish Certain Rules Respecting the Ownership of Low-Level Radioactive Waste, to Establish a Research, Development, and Demonstration Program for the Disposal of Radioactive Waste, and for Other Purposes.* 96th Cong., 2d sess., Washington, DC: GPO, 1980. 240p.

———. *West Valley Demonstration Project Act: Hearing, July 28, 1980, on H.R. 3193 and H.R. 6865, Bills to Authorize the Department of Energy to Carry Out a High-Level Liquid Nuclear Waste Management Demonstration Project at the Western New York Service Center in West Valley, New York, to Establish Procedures to Govern the Disposition of Other Nuclear Wastes at Such Sites and for Other Purposes.* 96th Cong., 2d sess., Washington, DC: GPO, 1980. 173p.

U.S. Congress. House. Committee on Interior and Insular Affairs. Subcommittee on Energy and the Environment. *Implementation of the Nuclear Waste Policy Act (Site Selection Program): Oversight Hearing, July 31, 1986.* 99th Cong., 2d sess., Washington, DC: GPO, 1987. 190p.

U.S. Congress. House. Committee on Science and Technology. Subcommittee on Investigations and Oversight. *West Valley [N.Y.] Cooperative Agreement: Hearing, July 9, 1981.* 97th Cong., 1st sess., Washington, DC: GPO, 1981. 371p.

U.S. Congress. Senate. Committee on Energy and Natural Resources. Subcommittee on Energy Research and Development. *Second Waste Repository Site Selection: Hearing, June 16, 1986, on the Second Waste Repository Site Selection under the Department of Energy's Office of Civilian Radioactive Waste Management.* 99th Cong., 2d sess., Washington, DC: GPO, 1986. 257p.

U.S. Department of Energy. *Environmental Assessment. Davis Canyon Site, Utah.* Washington, DC: Office of Civilian Radioactive Waste Management, 1986. 3 vols.

———. *Environmental Assessment: Deaf Smith County Site, Texas.* Washington, DC: Office of Civilian Radioactive Waste Management, 1986. 3 vols.

______. *Environmental Assessment: Overview, Davis Canyon Site, Utah.* Washington, DC: Office of Civilian Radioactive Waste Management, 1986. 38p.

______. *Environmental Assessment: Overview, Deaf Smith County Site, Texas.* Washington, DC: Office of Civilian Radioactive Waste Management, 1986. 36p.

______. *Environmental Assessment: Overview, Reference Repository Location, Hanford Site, Washington.* Washington, DC: Office of Civilian Radioactive Waste Management, 1986. 34p.

______. *Environmental Assessment: Overview, Richton Dome Site, Mississippi.* Washington, DC: Office of Civilian Radioactive Waste Management, 1986. 37p.

______. *Environmental Assessment: Overview, Yucca Mountain Site, Nevada, Research and Development Area, Nevada.* Washington, DC: Office of Civilian Radioactive Waste Management, 1986. 36p.

______. *Environmental Assessment: Reference Repository Location, Hanford Site, Washington.* Washington, DC: Office of Civilian Radioactive Waste Management, 1986. 3 vols.

______. *Environmental Assessment: Richton Dome Site, Mississippi.* Washington, DC: Office of Civilian Radioactive Waste Management, 1986. 3 vols.

______. *Environmental Assessment: Yucca Mountain Site, Nevada, Research and Development Area, Nevada.* Washington, DC: Office of Civilian Radioactive Waste Management, 1986. 3 vols.

______. *National Waste Storage Program: National Site Characterization and Selection Plan and Environmental Assessment.* Washington, DC, 1981.

Yow, Jesse L. Jr. "Air Injection to Modify Groundwater Flow." *Radioactive Waste Management* 2 (March 1982): 203–221.

Yu, Charley, William A. Jester, and Albert A. Jarrett. "Simultaneous Determination of Dispersion Coefficients and Retardation Factors for a Low Level Radioactive Waste Burial Site." *Radioactive Waste Management and the Nuclear Fuel Cycle* 4 (February 1984): 401–420.

Transportation

Barrett, L. H., and P. J. Grant. "Management, Treatment and Transportation of Nuclear Plant Waste: Solid, Liquid, and Gaseous." In Shyamal K. Majumdar and E. Willard Miller, eds., *Management of Radioactive Materials and Wastes: Issues and Progress* (Easton, PA: The Pennsylvania Academy of Science, 1985), 107–36. ISBN 0-9606670-4-0.

Church, Albert M., and Roger D. Norton. "Issues in Emergency Preparedness for Radiological Transportation Accidents." *Natural Resources Journal* 21 (October 1981): 757–771.

Cluett, Christopher, and Frederic A. Morris. "The Transportation of Radioactive Materials Through Urban Areas: Social Impacts and Policy Implications." *Southwestern Review of Management and Economics* 2 (Spring 1982): 207–221.

Garey, Robert B., and Richard C. Kearney. "State Policies on Nuclear Waste Transportation and Storage." *Texas Business Review* 54 (September–October 1980): 249–255.

Giglio, Sheila Bond. "Transportation of Spent Nuclear Fuel: The Need for a Flexible Regulatory System." *Boston College Environmental Affairs Law Review* 12 (Fall 1985): 51–101.

Levich, C., and James E. Martin. "The Dose to an Individual as a Result of the Incident-Free Transit of a Spent Fuel Transport Vehicle." *Radioactive Waste Management and the Nuclear Fuel Cycle* 7 (August 1986): 317–327.

Mallory, Charles W. "Transportation, Disposal and Conditioning of Low-Level Radioactive Waste." In Shyamal K. Majumdar and E. Willard Miller, eds., *Management of Radioactive Materials and Wastes: Issues and Progress* (Easton, PA: The Pennsylvania Academy of Science, 1985), 142–156. ISBN 0-9606670-4-0.

Schneider, K. J. "Radiation Dose Impacts Resulting from Variations in the Transportation-Related Activities in a System for Management of Spent Nuclear Fuel." *Nuclear Technology* 82 (July 1988): 106–113.

Organization for Economic Co-operation and Development. *Nuclear Energy Agency. Nuclear Legislation: Analytical Study: Regulations Governing the Transport of Radioactive Materials.* Paris: 1981, 201p.

U.S. Department of Transportation. Research and Special Programs Administration. *A Review of the Department of Transportation (DOT) Regulations for Transportation of Radioactive Materials.* Rev. ed. Washington, DC: GPO, 1983. 64p.

U.S. Department of Transportation. Research and Special Programs Administration. Materials Transportation Bureau. *A Review of the Department of Transportation (DOT) Regulations for Transportation of Radioactive Materials.* Washington, DC: GPO, 1980. 43p.

U.S. Congress. House. Committee on Energy and Commerce. Subcommittee on Telecommunications, Consumer Protection, and Finance. *Motor Carrier Safety: Transportation of Hazardous and Nuclear Materials: Hearing July 19, 1985.* 99th Cong., 1st sess., Washington, DC: GPO, 1986. 183p.

U.S. Congress. House. Committee on Public Works and Transportation. Subcommittee on Surface Transportation. *Transportation of Hazardous Materials Through City Streets: Hearing, May 27, 1980, on H.R. 792, to Amend the Hazardous Materials Transportation Act of 1974 to Prohibit the Transportation of Radioactive Materials in Densely Populated Areas.* 96th Cong., 2d sess., Washington, DC: GPO, 1980. 220p.

U.S. Congress. Senate. Committee on Governmental Affairs. Subcommittee on Energy, Nuclear Proliferation, and Government Processes. *Enforcement of Federal Regulations and Penalties for Shipments of Hazardous and Radioactive Materials: Hearing, May 9, 1984.* 98th Cong., 2d sess., Washington, DC: GPO, 1984. 169p.

Laws and Regulations

Barnaby, F. "Nuclear Weapons Free Zones" in *The Geography of Peace and War.* Edited by D. Pepper and A. Jenkins. Oxford, England: Basil Blackwell, 1985, pp. 163–177.

Bewers, J. M., and C. J. R. Garrett. "Analysis of the Issues Related to Sea Dumping of Radioactive Wastes." *Marine Policy* 11 (April 1987): 105–124.

Bowerman, B. S., R. E. Davis, and B. Siskind. *Document Review Regarding Hazardous Chemical Characteristics of Low-Level Waste.* Washington, DC: Division of Waste Management, Office of Nuclear Material Safety and Safeguards, U.S. Nuclear Regulatory Commission, 1986. 88p.

Choi, Yearn Hong. "Policy with Uneasy Implementation: U.S. Radioactive Waste Management." *Journal of Social, Political, and Economic Studies* 11 (Spring 1986): 83–91.

Draft Programmatic Environmental Impact Statement Related to Decontamination and Disposal of Radioactive Wastes Resulting from March 28, 1979, Accident Three Mile Island Nuclear Station, Unit 2, Docket No. 50-320, Metropolitan Edison Company, Jersey Central Power and Light Company, Pennsylvania Electric Company. NUREG, 0683. Washington, DC: Office of Nuclear Reactor Regulation, U.S. Nuclear Reactor Regulation, 1980. 1 vol.

Dunkelman, M. M., and others. *Plans and Schedules for Implementation of U.S. Nuclear Regulatory Commission Responsibilities under the Low-Level Radioactive Waste Policy Amendment Act of 1985 (P.L. 99-240).* Washington, DC: Division of Waste Management, Office of Nuclear Material Safety and Safeguards, U.S. Nuclear Regulatory Commission, 1986. 1 vol.

Graham, Stephen A. "The Nuclear Waste Policy Act of 1982: A Case Study in American Federalism [Including a Survey of the Attitudes of South Carolina Officials Toward Government Responsibility of Nuclear Waste]." *State Government* 57, no. 1 (1984): 7–12.

Malloy, Jane E. "1985 Department of Defense Authorization Act: Leaving Atomic Veterans at Ground Zero." *Valparaiso University Law Review* 20 (Spring 1986): 413–444.

Massachusetts. House of Representatives. Special Legislative Commission on Low-Level Radioactive Waste. *Report of the Special Commission Established to Make an Investigation and Study Relative to Low-Level Radioactive Waste, February 16, 1986.* Boston, MA: 1986. 95p.

Myers, Christopher W. *History, Structure and Institutional Overview of the Nuclear Waste Policy Act of 1982.* Santa Monica, CA: Rand, 1986. 33p.

New Jersey. Legislature. *Public Hearing on Senate Bill 3217 and Assembly Bill 3256 (Northeast Interstate Low-Level Radioactive Waste Management Compact) Held April 18, 1983, Trenton, New Jersey, Before Senate Energy and Environment Committee [and] Assembly Agriculture and Environment Committee.* Trenton: 1983. 50p.

Nichols, Teresa A., and Barbara H. Bink. "Special Report: The Reagan Administration, Congress, and Nuclear Power [Focuses on Licensing

Reform and Nuclear Waste Disposal]." *Public Utilities Fortnightly* 110 (November 11, 1982): 38–42.

Stensvaag, John-Mark, "Regulating Radioactive Air Emissions from Nuclear Generating Plants: A Primer for Attorneys, Decisionmakers, and Inventors." *Northwestern University Law Review* 78 (March 1983): 1–197.

U.S. Congress. House. Committee on Energy and Commerce. Subcommittee on Energy Conservation and Power. *DOE Radioactive Waste Repository Program: Hearings, August 1, 1985–May 1, 1986.* 99th Cong. Washington, DC: GPO, 1986. 1114p.

______. *DOE Regulation of Mixed Waste: Hearing Before the Subcommittee on Energy Conservation and Power and the Subcommittee on Commerce, Transportation, and Tourism of the Committee on Energy and Commerce, on H.R. 2009 and H.R. 2593, April 10, 1986.* 99th Cong., 2d sess., Washington, DC: 1986. 182p.

______. *Nuclear Waste Disposal Policy: Hearings, June 8 and 10, 1982, on H.R. 1993 [and Other Bills] to Provide for the Development of Facilities for Storage, Disposal and Reprocessing of Radioactive Waste and Spent Fuel, and for Other Purposes.* 97th Cong., 2d sess., Washington, DC: GPO, 1982. 603p.

______. Nuclear Waste Issues: *Hearings, June 11 and October 16, 1987, on H.R. 2888 and H.R. 2967, Bills to Suspend Certain Activities of the Department of Energy Under the Nuclear Waste Policy Act of 1982, to Establish a Nuclear Waste Policy Review Commission, an Office of the Nuclear Waste Negotiator, and for Other Purposes.* 100th Cong., 1st sess., Washington, DC: GPO, 1988. 327p.

______. *Price-Anderson Legislation: Hearing, July 17, 1986, on H.R. 2524 and H.R. 4394, Bills to Amend the Price-Anderson Provisions of the Atomic Energy Act of 1954 to Establish Liability and Indemnification for Nuclear Incidents Arising Out of Federal Storage, Disposal, or Related Transportation of High-Level Radioactive Waste and Spent Nuclear Fuel.* 99th Cong., 2d sess., Washington, DC: GPO, 1987. 344p.

U.S. Congress. House. Committee on Foreign Affairs. *United States-Japan Nuclear Cooperation Agreement: Hearings, December 16, 1987 and March 2, 1988.* 100th Cong., 1st sess., Washington, DC: GPO, 1988. 706p.

U.S. Congress. House. Committee on Interior and Insular Affairs. Subcommittee on Energy and the Environment. *Decommissioning of Nuclear Power Plants: Oversight Hearing, April 23, 1987.* 100th Cong., 1st sess., Washington, DC: GPO, 1988. 126p.

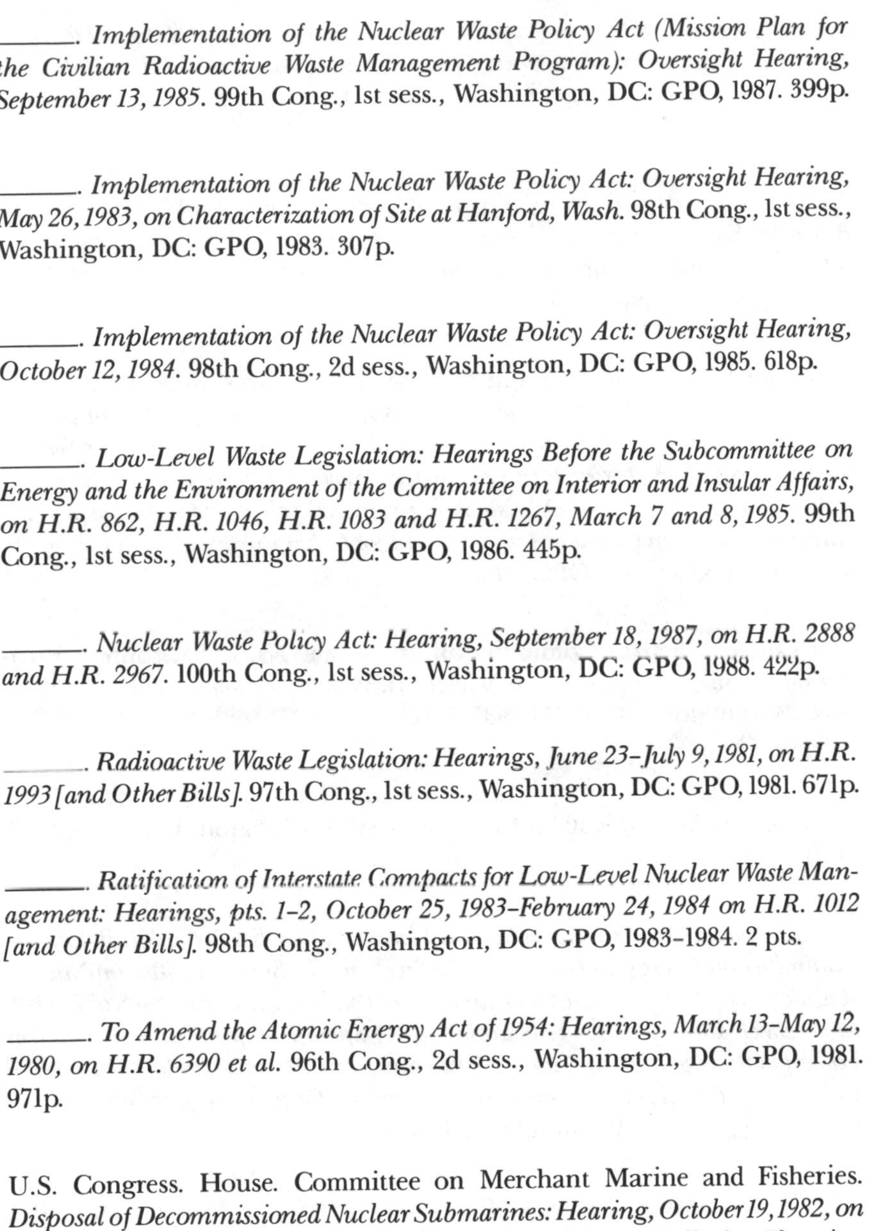

______. *Implementation of the Nuclear Waste Policy Act (Mission Plan for the Civilian Radioactive Waste Management Program): Oversight Hearing, September 13, 1985.* 99th Cong., 1st sess., Washington, DC: GPO, 1987. 399p.

______. *Implementation of the Nuclear Waste Policy Act: Oversight Hearing, May 26, 1983, on Characterization of Site at Hanford, Wash.* 98th Cong., 1st sess., Washington, DC: GPO, 1983. 307p.

______. *Implementation of the Nuclear Waste Policy Act: Oversight Hearing, October 12, 1984.* 98th Cong., 2d sess., Washington, DC: GPO, 1985. 618p.

______. *Low-Level Waste Legislation: Hearings Before the Subcommittee on Energy and the Environment of the Committee on Interior and Insular Affairs, on H.R. 862, H.R. 1046, H.R. 1083 and H.R. 1267, March 7 and 8, 1985.* 99th Cong., 1st sess., Washington, DC: GPO, 1986. 445p.

______. *Nuclear Waste Policy Act: Hearing, September 18, 1987, on H.R. 2888 and H.R. 2967.* 100th Cong., 1st sess., Washington, DC: GPO, 1988. 422p.

______. *Radioactive Waste Legislation: Hearings, June 23–July 9, 1981, on H.R. 1993 [and Other Bills].* 97th Cong., 1st sess., Washington, DC: GPO, 1981. 671p.

______. *Ratification of Interstate Compacts for Low-Level Nuclear Waste Management: Hearings, pts. 1–2, October 25, 1983–February 24, 1984 on H.R. 1012 [and Other Bills].* 98th Cong., Washington, DC: GPO, 1983–1984. 2 pts.

______. *To Amend the Atomic Energy Act of 1954: Hearings, March 13–May 12, 1980, on H.R. 6390 et al.* 96th Cong., 2d sess., Washington, DC: GPO, 1981. 971p.

U.S. Congress. House. Committee on Merchant Marine and Fisheries. *Disposal of Decommissioned Nuclear Submarines: Hearing, October 19, 1982, on Oversight of the Ocean Dumping Act and National Ocean Pollution Planning Act and the Disposal of Defueled Decommissioned Nuclear Submarines.* 97th Cong., 2d sess., Washington, DC: GPO, 1983. 166p.

U.S. Congress. House. Committee on Science and Technology. Subcommittee on Energy Research and Production. *H.R. 5016, Nuclear Waste Management Comprehensive Legislation—Bouquard/Lujan Proposal: Hearings, June 17–October 7, 1981.* 97th Cong., 1st sess., Washington, DC: GPO, 1982. 670p.

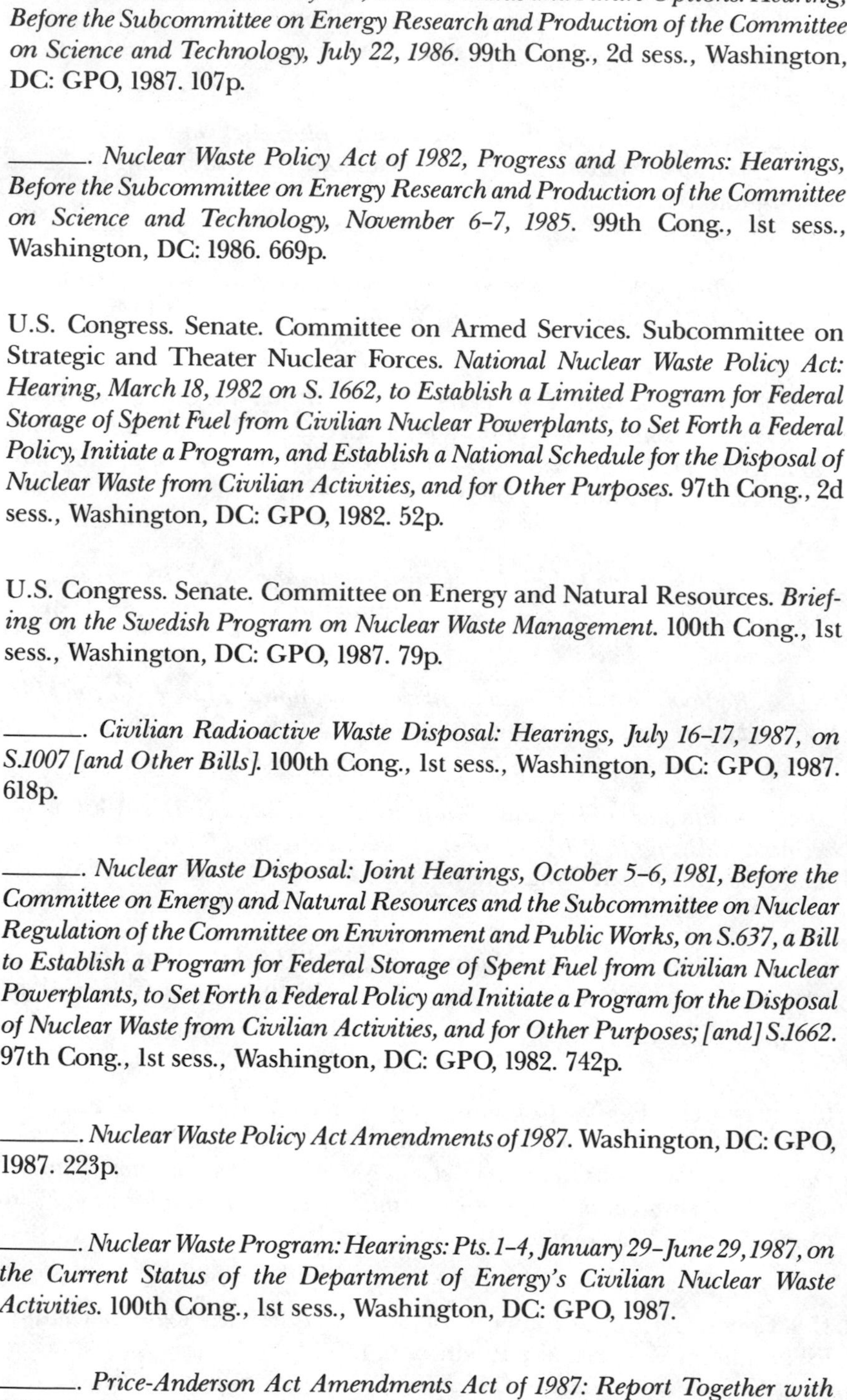

———. *Nuclear Waste Policy Act, Current Status and Future Options: Hearing, Before the Subcommittee on Energy Research and Production of the Committee on Science and Technology, July 22, 1986.* 99th Cong., 2d sess., Washington, DC: GPO, 1987. 107p.

———. *Nuclear Waste Policy Act of 1982, Progress and Problems: Hearings, Before the Subcommittee on Energy Research and Production of the Committee on Science and Technology, November 6–7, 1985.* 99th Cong., 1st sess., Washington, DC: 1986. 669p.

U.S. Congress. Senate. Committee on Armed Services. Subcommittee on Strategic and Theater Nuclear Forces. *National Nuclear Waste Policy Act: Hearing, March 18, 1982 on S. 1662, to Establish a Limited Program for Federal Storage of Spent Fuel from Civilian Nuclear Powerplants, to Set Forth a Federal Policy, Initiate a Program, and Establish a National Schedule for the Disposal of Nuclear Waste from Civilian Activities, and for Other Purposes.* 97th Cong., 2d sess., Washington, DC: GPO, 1982. 52p.

U.S. Congress. Senate. Committee on Energy and Natural Resources. *Briefing on the Swedish Program on Nuclear Waste Management.* 100th Cong., 1st sess., Washington, DC: GPO, 1987. 79p.

———. *Civilian Radioactive Waste Disposal: Hearings, July 16–17, 1987, on S.1007 [and Other Bills].* 100th Cong., 1st sess., Washington, DC: GPO, 1987. 618p.

———. *Nuclear Waste Disposal: Joint Hearings, October 5–6, 1981, Before the Committee on Energy and Natural Resources and the Subcommittee on Nuclear Regulation of the Committee on Environment and Public Works, on S.637, a Bill to Establish a Program for Federal Storage of Spent Fuel from Civilian Nuclear Powerplants, to Set Forth a Federal Policy and Initiate a Program for the Disposal of Nuclear Waste from Civilian Activities, and for Other Purposes; [and] S.1662.* 97th Cong., 1st sess., Washington, DC: GPO, 1982. 742p.

———. *Nuclear Waste Policy Act Amendments of 1987.* Washington, DC: GPO, 1987. 223p.

———. *Nuclear Waste Program: Hearings: Pts. 1–4, January 29–June 29, 1987, on the Current Status of the Department of Energy's Civilian Nuclear Waste Activities.* 100th Cong., 1st sess., Washington, DC: GPO, 1987.

———. *Price-Anderson Act Amendments Act of 1987: Report Together with Minority Views (to Accompany S.748).* Washington, DC: GPO, 1987. 91p.

______. *Price-Anderson Act Amendments Act of 1986: Report Together with Additional and Minority Views (to Accompany S.1225).* Washington, DC: GPO, 1987. 107p.

U.S. Congress. Senate. Committee on Energy and Natural Resources. Subcommittee on Energy Research and Development. *Low-Level Radioactive Waste Disposal: Joint Hearing Before the Subcommittee on Energy and Natural Resources and the Subcommittee on Nuclear Regulation of the Committee on Environment and Public Works, on S.1517 and S.1578, October 8, 1985.* 99th Cong., 1st sess., Washington, DC: GPO, 1986. 681p.

U.S. Congress. Senate. Committee on Environment and Public Works. Subcommittee on Nuclear Regulation. *The Economic Implications of Locating a Nuclear Waste Repository in Texas: Hearing, February 11, 1985.* 99th Cong., 1st sess., Washington, DC: GPO, 1985. 172p.

______. *High-Level Nuclear Waste Issues: Hearings, April 23–June 18, 1987.* 100th Cong., 1st sess., Washington, DC: GPO, 1987. 828p.

______. *Mixed Radioactive and Hazardous Waste Disposal Issues, Joint Hearing Before the Subcommittees on Nuclear Regulation and Environmental Pollution of the Committee on Environment and Public Works on S.892, March 25, 1986.* 99th Cong., 2d sess., Washington, DC: GPO, 1986. 198p.

______. *National Nuclear Waste Policy Act of 1981: Hearings, October 31 and November 9, 1981, on S.1662, a Bill to Establish a Limited Program for Federal Storage of Spent Fuel from Civilian Nuclear Powerplants, to Set Forth A Federal Policy, Initiate a Program and Establish a National Schedule for the Disposal of Nuclear Waste from Civilian Activities, and for Other Purposes.* 97th Cong., 1st sess., Washington, DC: GPO, 1982. 274p.

U.S. Congress. Senate. Committee on Governmental Affairs. *Environmental Issues at Department of Energy Nuclear Facilities: Hearings, March 17, 1987.* 100th Cong., 1st sess., Washington, DC: GPO, 1987. 182p.

U.S. Congress. Senate. Committee on the Judiciary. *Central Interstate Low-Level Radioactive Waste Compact: Hearing, October 21, 1983, on S.1581, a Bill Granting the Consent of Congress to the Central Interstate Low-Level Radioactive Waste Compact.* 98th Cong., 1st sess., Washington, DC: GPO, 1984. 99p.

______. *Management Compacts on Low-Level Radioactive Waste: Hearing, March 8, 1985, on S.44 [and Other Bills].* 99th Cong., 1st sess., Washington, DC: GPO, 1985. 246p.

———. *Rocky Mountain Low-Level Radioactive Waste Compact: Hearing, January 12, 1984 on S.1991*. 98th Cong., 2d sess., Washington, DC: GPO, 1984. 189p.

———. *Status of Interstate Compacts for the Disposal of Low-Level Radioactive Waste: Hearing, March 2, 1983*. 98th Cong., 1st sess., Washington, DC: GPO, 1983. 252p.

U.S. Department of Energy. *In the Matter of Proposed Rulemaking on the Storage and Disposal of Nuclear Waste (Waste Confidence Rulemaking): PR-50, 51 (44FRG1372)*. Washington, DC: 1980.

U.S. Department of Energy. Office of Civilian Radioactive Waste Management. *OCRWM Mission Plan Amendment*. Washington, DC: 1987. 517p.

U.S. General Accounting Office. *The Nuclear Waste Policy Act, 1984. Implementation Status, Progress, and Problems: Report to the Congress by the Comptroller General of the United States*. Washington, DC: U.S. General Accounting Office, 1986. 124p.

U.S. Laws, Statutes, etc. *Nuclear Regulatory Legislation, through the 97th Congress, 2d Session*. 98th Cong., 1st sess., Washington, DC: GPO, 1983. 282p.

Selected Journal Titles

The journals listed below publish articles on many aspects of the nuclear industry. Because the problems of the nuclear industry have only become important in recent years, new journals are continually appearing in print. For new journals and additional information please consult: *Ulrich's International Periodicals Directory* 1989–90. 28th edition. New York: R. R. Bowker Company, 1989. 3 vols.

Sample Entry

Journal Title

1. Editor
2. Year first published
3. Frequency of publication
4. Code

5. Special features
6. Address of publisher

Journals

Across the Board

1. Howard Muson
2. 1939
3. Monthly
4. —
5. Adv., charts, illus., index
6. Conference Board Inc.
   845 Third Avenue
   New York, NY 10022

American Journal of Science

1. Editorial Board
2. 1818
3. Monthly
4. ISSN 0002-9599
5. Bk. rev., bibl., illus., index
6. Kline Geological Laboratory
   American Journal of Science
   Box 6666
   Yale Station, New Haven, CT 06511

Atmospheric Environment

1. James P. Lodge, Jr.
2. 1967
3. Monthly
4. ISSN 0004-6981
5. Adv., bk. rev., charts, illus.
6. Pergamon Press, Inc.
   Journals Division
   Maxwell House
   Fairview Park
   Elmsford, NY 10523

Atom

1. Wendy Peters
2. 1956

3. Monthly
4. UK ISSN 0004-7015
5. Adv., bk. rev., charts, index
6. United Kingdom Atomic Energy Authority
   Information Services Branch
   11 Charles 2nd Street
   London SW1Y 4QP
   England

Boston College Environmental Affairs Law Review

1. Editorial Board
2. 1971
3. Quarterly
4. ISSN 0190-7034
5. Adv., bk. rev., bibl., charts
6. Boston College
   School of Law
   885 Centre Street
   Newton Centre, MA 02159

Bulletin of the Atomic Scientists

1. Len Ackland
2. 1945
3. 10 issues per year
4. ISSN 0096-3402
5. Adv., bk. rev., illus., index
6. Educational Foundation for Nuclear Science
   5801 S. Kenwood Avenue
   Chicago, IL 60637

Chemical Engineering Progress

1. Attilio Bisio
2. 1947
3. Monthly
4. ISSN 0360-7275
5. Adv., abstr., charts, illus., tr. lit., index
6. American Institute of Chemical Engineers
   345 East 47th Street
   New York, NY 10017

Corrosion Science

1. J.D. Scully
2. 1961
3. Monthly
4. ISSN 0010-938X
5. Adv., charts, illus., index
6. Pergamon Press, Inc.
   Journals Division
   Maxwell House
   Fairview Park
   Elmsford, NY 10523

Department of State Bulletin

1. Phyllis A. Young
2. 1939
3. Monthly
4. ISSN 0041-7610
5. Bibl., charts, illus., index
6. U.S. Department of State
   Bureau of Public Affairs
   Washington, DC 20502

Disarmament (UN)

1. —
2. 1978
3. Irreg., 2–3/year
4. —
5. Bibl., charts
6. United Nations Publications
   Sales Section
   Room DC2-853
   New York, NY 10017

EPRI Journal

1. Brent Barker
2. 1976
3. 10 issues per year
4. ISSN 0362-3416
5. Index

6. Electric Power Research Institute
   Box 10412
   3412 Hillview Avenue
   Palo Alto, CA 94304

Ecology Law Quarterly

1. Beverly Alexander
2. 1971
3. Quarterly
4. ISSN 0046-1121
5. Adv., bk. rev., bibl., index
6. University of California Press
   Journals Division
   2120 Berkeley Way
   Berkeley, CA 94720

Editorial Research Reports

1. Hoyt Gimlin
2. 1923
3. 4 times a month
4. ISSN 0013-0958
5. Bk. rev., charts, index
6. Congressional Quarterly Inc.
   1414 22nd Street, NW
   Washington, DC 20037

Energy Policy

1. —
2. 1973
3. Bimonthly
4. UK ISSN 0301-4215
5. Adv., bk. rev., charts, illus., stat., index
6. Butterworth Scientific Ltd.
   P.O. Box 63
   Westbury House
   Bury Street
   Guilford, Surrey GU2 5BH
   England

Environment

1. Jane Scully
2. 1958

3. Monthly (except Jan.-Feb., July-Aug. combined)
4. ISSN 0013-9157
5. Adv., bk. rev., bibl., charts, illus., index, cum. index
6. Heldref Publications
   4000 Albemarle Street, NW
   Washington, DC 20016

Environmental Action

1. Editorial Board
2. 1970
3. Bimonthly
4. ISSN 0013-922X
5. Adv., bk. rev., film rev., illus., index
6. Environmental Action, Inc.
   1525 New Hampshire Avenue, NW
   Washington, DC 20036

Environmental Conservation

1. Nicholas Polunin
2. 1974
3. Quarterly
4. ISSN 0376-8929
5. Bk. rev.
6. Elsevier Sequoia S.A.
   Box 564
   CH-1001 Lausanne 1
   Switzerland

Environmental Research

1. I.J. Selikoff
2. 1967
3. Bimonthly
4. ISSN 0013-9351
5. Adv., bk. rev., illus., index
6. Academic Press, Inc.
   Journal Division
   1250 Sixth Avenue
   San Diego, CA 92101

Environmental Science & Technology

1. Russell F. Christman
2. 1967

3. Monthly
4. ISSN 0013-936X
5. Adv., bk. rev., abstr., bibl., charts, illus., stat., index
6. American Chemical Society
   1155 16th Street, NW
   Washington, DC 20036

Environmentalist

1. John F. Potter
2. 1980
3. Quarterly
4. UK ISSN 0251-1088
5. Adv.
6. Science and Technology Letters
   12 Clarence Road
   Kew, Surrey TW9 3NL
   England

Foreign Affairs

1. William G. Hyland
2. 1922
3. 5 times a year
4. ISSN 0015-7120
5. Adv., bk. rev., index
6. Council on Foreign Relations, Inc.
   58 E. 68th Street
   New York, NY 10021

Foreign Policy

1. Charles Maynes
2. 1970
3. Quarterly
4. ISSN 0015-7228
5. Bk. rev., index every 2 years
6. Carnegie Endowment for International Peace
   11 Dupont Circle NW
   Suite 900
   Washington, DC 20036

Fusion Technology

1. George H. Miley
2. 1981

3. Bimonthly
4. ISSN 0748-1896
5. Bk. rev., charts
6. American Nuclear Society
   555 N. Kensington Avenue
   La Grange Park, IL 60525

Geophysics

1. —
2. 1936
3. Monthly
4. ISSN 0016-8033
5. Adv., bk. rev., abstr., illus., pat., index
6. Society of Exploration Geophysicists
   Box 702740
   Tulsa, OK 74710-2740

Glass Technology

1. B. E. Moody
2. 1960
3. Bimonthly
4. ISSN 0017-1050
5. Adv., bk. rev., abstr., index
6. Society of Glass Technology
   Thornton, 20 Hallam Gate Road
   Sheffield S10 5BT
   England

Health Physics

1. Dr. Genevieve S. Roessler
2. 1958
3. Monthly
4. ISSN 0017-9078
5. Adv., charts, illus., stat., index
6. Pergamon Press, Inc.
   Journals Division
   Maxwell House
   Fairview Park
   Elmsford, NY 10523

International Affairs (London)

1. J. Roper
2. 1922
3. 4 times a year
4. ISSN 0020-5850
5. Adv., bk. rev., index
6. Butterworth Scientific Ltd.
   P.O. Box 63
   Westbury House, Bury Street
   Guilford, Surrey GU2 5BH
   England

International Journal on World Peace

1. Panos D. Bardis
2. 1984
3. Quarterly
4. ISSN 0742-3640
5. Adv., bk. rev.
6. Professors World Peace Academy
   G.P.O. 1311
   New York, NY 10116

International Security

1. Ashton B. Carter and Steven E. Miller
2. 1976
3. Quarterly
4. ISSN 0162-2889
5. Adv.
6. MIT Press
   55 Hayward Street
   Cambridge, MA 02142

Journal of Applied Ecology

1. T.M. Roberts and W. Block
2. 1964
3. 3 times a year
4. ISSN 0021-8901
5. Adv., bk. rev., bibl., charts, illus., index
6. Blackwell Scientific Publications Ltd.
   Osney Mead
   Oxford OX2 OEL
   England

Journal of Energy Law and Policy

1. Editorial Board
2. 1980
3. Twice annually
4. ISSN 0275-9926
5. Bk. rev., index
6. University of Utah
   College of Law
   Salt Lake City, UT 84112

Journal of Environmental Engineering

1. —
2. 1956
3. Bimonthly
4. ISSN 0733-9372
5. —
6. American Society of Civil Engineers
   345 East 47th Street
   New York, NY 10017

Journal of Environmental Quality

1. T. J. Logan
2. 1972
3. Quarterly
4. ISSN 0047-2425
5. Adv., bk. rev., bibl., charts, illus., stat.
6. American Society of Agronomy, Inc.
   677 South Segoe Road
   Madison, WI 53711

Journal of Environmental Radioactivity

1. M. S. Baxter
2. —
3. 4 issues per year
4. UK ISSN 0265-931X
5. —
6. Elsevier Applied Science Publishers, Ltd.
   22 Rippleside Commercial Estate
   Barking Essex
   England IG11 OSA

Journal of Environmental Sciences

1. Janet A. Ehmann
2. 1959
3. Bimonthly
4. ISSN 0022-0906
5. Adv., bk. rev., abstr., illus.
6. Institute of Environmental Sciences
   940 E. Northwest Highway
   Mt. Prospect, IL 60056

Journal of Geotechnical Engineering

1. —
2. 1956
3. Monthly
4. ISSN 0733-9410
5. —
6. American Society of Civil Engineers
   345 East 47th Street
   New York, NY 10017

Journal of Hazardous Materials

1. G. F. Bennett and R. F. Griffiths
2. 1975
3. 6 issues per year
4. ISSN 0304-3894
5. Adv., bk. rev., index
6. Elsevier Science Publishers B.V.
   P.O. Box 211
   1000 AE Amsterdam
   Netherlands

Journal of Health Politics, Policy and Law

1. Lawrence D. Brown
2. 1976
3. Quarterly
4. ISSN 0361-6878
5. Adv., bk. rev.
6. Duke University Press
   6697 College Station
   Durham, NC 27708

Journal of Peace Research

1. Nils Petter Gleditsch
2. 1964
3. Quarterly
4. ISSN 0022-3433
5. Adv., bk. rev., charts, illus., stat., index
6. Publications Expediting Inc.
   200 Meacham Avenue
   Elmont, NY 11003

Journal of Policy Analysis Management

1. David L. Weimer
2. 1981
3. Quarterly
4. ISSN 0276-8739
5. Adv., index
6. John Wiley & Sons, Inc.
   605 Third Avenue
   New York, NY 10016

Long Range Planning

1. Bernard Taylor
2. 1968
3. Bimonthly
4. ISSN 0024-6301
5. Adv., bk. rev., charts, stat.
6. Pergamon Press, Inc.
   Journals Division
   Maxwell House
   Fairview Park
   Elmsford, NY 10523

Marine Geotechnology

1. Ronald Chaney
2. 1975
3. Quarterly
4. ISSN 0360-8867
5. Adv., bk. rev., abstr., charts, illus., stat., index
6. Taylor & Frances, Crane, Russak
   3 East 44th Street
   New York, NY 10017

Marine Policy

1. Editorial Board
2. 1977
3. Quarterly
4. UK ISSN 0308-597X
5. Bk. rev., abstr., illus.
6. Butterworth Scientific Ltd.
   P.O. Box 63
   Westbury House
   Bury Street
   Guilford, Surrey GU2 5BH
   England

Natural Resources Journal

1. Albert E. Utton
2. 1961
3. Quarterly
4. ISSN 0028-0739
5. Adv., bk. rev., charts, index, cum. index every 10 years
6. University of New Mexico
   School of Law
   1117 Stanford NE
   Albuquerque, NM 87131

Nuclear and Chemical Waste Management

1. A. Moghissi
2. 1980
3. Quarterly
4. ISSN 0191-815X
5. Adv., illus.
6. Pergamon Press, Inc.
   Journals Division
   Maxwell House
   Fairview Park
   Elmsford, NY 10523

Nuclear Energy

1. J. L. Head
2. 1962
3. 6 issues per year
4. UK ISSN 1040-4067
5. Adv., bk. rev., charts, illus., index

6. Thomas Telford Ltd.
   1-7 Great George Street
   Westminster, London SWIP 3AA
   England

Nuclear Engineering International

1. R. Masters
2. 1956
3. Monthly
4. ISSN 0029-5507
5. Adv., bk. rev., charts, illus., tr. lit., index
6. Electrical-Electronic Press
   Quadrant House
   The Quadrant
   Sutton, Surrey SM2 5AS
   England

Nuclear News

1. Jon Payne
2. 1959
3. Monthly (plus 3 special issues)
4. ISSN 0029-5574
5. Adv., bk. rev., charts, illus., tr. lit.
6. American Nuclear Society
   555 N. Kensington Avenue
   La Grange Park, IL 60525

Nuclear Safety

1. —
2. 1959
3. Bimonthly
4. ISSN 0029-5604
5. Bk. rev., charts, illus., index
6. U.S. Department of Energy
   Office of Scientific and Technical Information
   Box 62
   Oak Ridge, TN 37831

Nuclear Technology

1. Roy G. Post
2. 1965
3. Monthly

4. ISSN 0029-5450
5. Bk. rev., charts, illus., stat., index
6. American Nuclear Society
   555 N. Kensington Avenue
   La Grange Park, IL 60525

Ocean Development and International Law

1. Daniel S. Cheever
2. 1973
3. Bimonthly
4. ISSN 0090-8320
5. Adv., bk. rev., abstr., charts, stat., index
6. Taylor & Francis, Crane, Russak
   3 East 44th Street
   New York, NY 10017

Oceanus

1. Paul R. Ryan
2. 1952
3. Quarterly
4. ISSN 0029-8182
5. Adv., bk. rev., charts, illus., index
6. Woods Hole Oceanographic Institution
   Box 6419
   Syracuse, NY 13217

Parameters

1. Lloyd J. Matthews
2. 1971
3. Quarterly
4. ISSN 0031-1723
5. Bk. rev., charts, illus.
6. U.S. Army War College
   Carlisle Barracks, PA 17013

Peace and Change

1. Robert D. Schulzinger and Dennis Carey
2. 1972
3. Quarterly
4. ISSN 0149-0508
5. Adv., bk. rev.

6. Consortium on Peace Research, Education, and Development
   c/o Robert D. Schulzinger
   History Department
   University of Colorado
   Boulder, CO 80309

Power Engineering

1. Robert Smock
2. 1896
3. Monthly
4. ISSN 0032-5961
5. Bk. rev., abstr., bibl., charts, illus., stat., index
6. Penn Well Publishing Company
   1421 Sheridan Road
   Tulsa, OK 74112

Radioactive Waste Management (Oak Ridge)

1. —
2. 1981
3. Twice monthly
4. ISSN 0275-3707
5. —
6. U.S. Department of Energy
   Office of Scientific and Technical Information
   Box 62
   Oak Ridge, TN 37831

Radioactive Waste Management and the Nuclear Fuel Cycle
Formerly: *Radioactive Waste Management*

1. A. M. Platt
2. 1980
3. 8 issues per year (in 2 vols., 4 nos. per vol.)
4. ISSN 0739-5876
5. —
6. Harwood Academic Publishers
   50 West 23rd Street
   New York, NY 10010

Science

1. Daniel Koshland
2. 1880
3. Weekly (4 volumes per year)

4. ISSN 0036-8075
5. Adv., bk. rev., abstr., bibl., illus.
6. American Association for the Advancement of Science
   1333 H Street, NW
   Washington, DC 20005

Science of the Total Environment

1. E. I. Hamilton
2. 1972
3. 21 times per year
4. NE ISSN 0048-9697
5. Adv., bk. rev., charts, illus., index
6. Elsevier Science Publishers B.V.
   Box 211
   1000 AE Amsterdam
   Netherlands

Science Technology and Human Values

1. Marcel Chatkowski
2. 1972
3. Quarterly
4. ISSN 0162-2439
5. Bibl., index
6. John Wiley & Sons, Inc.
   605 Third Avenue
   New York, NY 10158

Scientific American

1. Jonathan Peil
2. 1845
3. Monthly
4. ISSN 0036-8733
5. Adv., bk. rev., illus, index
6. Scientific American Inc.
   415 Madison Avenue
   New York, NY 10017

Sea Technology

1. David M. Graham
2. 1960
3. Monthly
4. ISSN 0093-3651

5. Adv., charts, illus., tr. lit, index
6. Compass Publications, Inc.
   (Arlington) Suite 1000
   117 N. 19th Street
   Arlington, VA 22209

Social Problems

1. James Orcutt
2. 1953
3. 5 times per year
4. ISSN 0037-7791
5. Adv., bk. rev., bibl., charts, index, cum. index
6. University of California Press
   2120 Berkeley Way
   Berkeley, CA 94720

Strategic Review

1. Walter F. Hahn
2. 1973
3. Quarterly
4. ISSN 0091-6846
5. Bk. rev.
6. United States Strategic Institute
   265 Winter Street
   Waltham, MA 02154

Technology Review

1. John Mattill
2. 1899
3. 8 issues per year
4. ISSN 0040-1692
5. Adv., bk. rev., illus., index
6. Massachusetts Institute of Technology
   Alumni Association, 10-40
   Cambridge, MA 02139

World Policy

1. Sherle Schwenninger
2. 1983
3. Quarterly

4. ISSN 0740-2775
5. Adv., bk. rev., charts
6. World Policy Institute
777 United Nations Plaza
New York, NY 10017

6

Films, Filmstrips, and Videocassettes

THE FILMS AND FILMSTRIPS ON THE NUCLEAR INDUSTRY provide a wide spectrum of information, from its peaceful uses in providing energy to the devastation of an atomic bomb. The selection begins with a series of films on atomic physics providing the theoretical background to the specific aspects of nuclear applications. The graphic presentation provides information more vividly than the written word. The public must reach a critical decision within a short time of the place of nuclear energy in modern-day society. A film presentation may aid in reaching a satisfactory solution. Although the films vary greatly in date of production, all films included remain technically and scientifically accurate.

The following sources list films and filmstrips in English.

Educational Film/Video Locator of the Consortium of University Film Centers and R. R. Bowker. 3d ed. New York: R. R. Bowker Company, 1986, 2 vols. 3115p.

Films in the Sciences: Reviews and Recommendations. Washington, DC: American Association for the Advancement of Science, 1980. 172p.

Index to Environmental Studies (Multimedia). University Park, Los Angeles, CA: National Information Center for Educational Media (NICEM), University of Southern California, 1977. 1113p.

The Video Source Book. 9th ed. Syosset, NY: The National Video Clearinghouse, Inc., 1986. 2224p. and supplements.

Atomic Physics

Atomic Energy for Space

Handel Film Corp.
8730 Sunset Boulevard
Los Angeles, CA 90069
Color, 17 minutes, sound, 16mm, 1966.

Animation explains the fission process, the nuclear power reactor, the isotopic space generator, and SNAP devices.

Atomic Energy—Inside the Atom

Encyclopedia Britannica Educational Corp.
425 N. Michigan Avenue
Chicago, IL 60611
Color/black and white, 13 minutes, sound, 16mm, 1961, 2d ed. 1982

Describes the atom, its parts, and the energy of the atom. Discusses stability and instability of the nucleus in relation to radiation.

Atomic Physics, Part 1—The Atomic Theory

Universal Education and Visual Arts
100 Universal City Plaza
Universal City, CA 91608
Black and white, 9 minutes, sound, 16mm, 1948.

Presents the basic theory proposed by Dalton in 1808 and outlines program of atomic study in the nineteenth century, including Faraday's electrolyses and Mendeleev's periodic table.

Atomic Physics, Part 2—Rays from Atoms

Universal Education and Visual Arts
100 Universal City Plaza
Universal City, CA 91608
Black and white, 11 minutes, sound, 16mm, 1948.

Demonstrates early work with cathode rays and discovery of the electron. How positive rays were discovered and their nature and the work of Roentgen with X-rays.

Atomic Physics, Part 3—The Nuclear Structure of the Atom

Universal Education and Visual Arts
100 Universal City Plaza
Universal City, CA 91608
Black and white, 19 minutes, sound, 16mm, 1948.

Shows alpha, beta, and gamma rays as examples of radioactivity. Discusses the uranium family and explains the relationship between atomic weight and atomic number.

Atomic Physics, Part 4—Atom Smashing and the Discovery of the Neutron

Universal Education and Visual Arts
100 Universal City Plaza
Universal City, CA 91608
Black and white, 22 minutes, sound, 16mm, 1948.

Explains the work of the Curies and James Chadwick in the discovery of the neutron. Discusses the splitting of the lithium atom by Cockcroft and Walton. Einstein epxlains his theory of mass and energy.

Atomic Physics, Part 5—Uranium Fission: Atomic Energy

Universal Education and Visual Arts
100 Universal City Plaza
Universal City, CA 91608
Black and white, 24 minutes, sound, 16mm, 1948.

Shows the development of the cyclotron, nuclear bombardment, nuclear fission, and chain reaction. Explores peacetime use of atomic energy.

Nuclear Technology

Atomic Energy: Inside the Atom (2nd ed.)

Britannica Films
425 North Michigan Avenue
Chicago, IL 60611
Color, 14 minutes, sound, Beta, VHS, 3/4″ U-matic cassette, 1982.

The film shows there is energy inside the atom and the ways in which the radioactive substances can be used.

Fusion: The Ultimate Fire

BFA Educational Media
P.O. Box 1795
2211 Michigan Avenue
Santa Monica, CA 90404
Color, 14 minutes, sound, 16mm, 1976.

This film presents the technical aspects of the development of fusion as a power source.

Nuclear Magnetic Resonance

University of California Extensive Media Center
2223 Fulton Street
Berkeley, CA 94760
Color, 27 minutes, sound, 16mm, 1966.

Introduces the theory of nuclear magnetic resonance through the use of animated graphs. Explains how NMR varies with radio frequencies and magnetic fields. Gives chemical shift, relative areas, and spin splotting methods.

Nuclear Reactor

CRM/McGraw-Hill Films
110 Fifteenth Street
Del Mar, CA 92014
Black and white, 10 minutes, sound, 16mm, 1954.

Describes the structure of an atomic pile and the method of control of the radioactivity of the pile by means of mixing the fissionable materials with lighter elements and inserting rod absorbers at various depths. Uses antiquated diagrams and charts.

Radioactive Dating

Coronet Films and Associates
121 North West Crystal Street
Crystal River, FL 32629
Color, 13 minutes, sound, Beta, VHS, 3/4″ U-matic cassette, other formats availabe from the distributor, 1981.

Radioactive dating is described and explains radioactive decay and half-life. Shows how radioactive substances such as carbon 14 and potassium 40 can be used by scientists to date materials.

Nuclear Analysis

Nuclear Fingerprinting of Ancient Pottery

United States Atomic Energy Commission
Division of Public Information
Audio-Visual Branch
Washington, DC 20545
Color, 20 minutes, sound, 16mm, 1970.

Shows the techniques and research used to identify pottery through chemical and nuclear analysis. Examines the processes in dating pottery and in determining its place of origin. Animation and documentary sequences are used.

Radiation

Radiation

National Film Board of Canada
1251 Avenue of the Americas
New York, NY 10020
Color, 27 minutes, sound, 16mm, 1959.

Introduces sunlight as a type of radiation and points out that just as there is a color spectrum, there is also a radiation spectrum. Illustrates several examples of radiation, such as sunlight and radio waves. Shows effect of sunlight on plants as it stunts, deforms, cripples, and causes mutations.

Radiation Biology

American Institute of Biological Sciences
1401 Wilson Boulevard
Arlington, VA 22209
Color, 27 minutes, sound, 16mm, 1961.

Illustrates the nature, detection, effects and use of radioactivity. Demonstrates the use of detection equipment and use of radiation by the ecologist.

Radiation Effects on Farm Animals

United States Department of Agriculture
Motion Picture Service
South Building, Room 1850
Washington, DC 20230
Color, 14 minutes, sound, 16mm, 1964.

Deals with the immediate effects of high radiation exposure received by farm animals over a short period of time. Shows symptoms of the pathological changes. Discusses the amount and rate of radiation exposure and the effects on the health of the animals' biological system. Animation is used.

Radiation Sources

Telstar Productions Inc.
366 North Prior Avenue
St. Paul, MN 55104
Black and white, 30 minutes, sound, Beta, VHS, 1/2" open reel (EIAJ), 3/4" U-matic cassette, 1967.

Discusses sources of radiation such as fallout, cosmic radiation, and isotope formation. Gives the physical and chemical makeup of gamma rays and alpha and beta particles.

Radioactivity

CRM/McGraw-Hill Films
110 Fifteenth St.
Del Mar, CA 92014
Color/black and white, 13 minutes, sound, 16mm, 1962.

Uses a series of demonstrations to discuss basic aspects of radioactivity. Describes the characteristics of alpha, beta, and gamma radiation. Utilizes the Wilson cloud chamber and radiation scale in detecting and measuring radiation in terms of mass and change. Shows the mass spectrograph and oxygen within an evacuated tube. Stresses the nucleus of the atom as the source of radiation.

Radioactivity

Encyclopaedia Britannica Educational Corp.
425 N. Michigan Ave.
Chicago, IL 60611
Black and white, 30 minutes, sound, 16mm, 1957.

Discusses the discovery of radioactivity by Becquerel and the ionization of alpha, beta, and gamma rays. Demonstrates the principles of the Wilson cloud chamber.

Radioactivity

Michigan Media
University of Michigan
400 Fourth Street
Ann Arbor, MI 48109
Color, 29 minutes, sound, 3/4" U-matic cassette, other formats available from the distributor, 1979.

Shows how scientists and engineers control radiation in power reactors. Also describes how radioactivity works for and against us.

Radioactivity Measurements: Laboratory

Encyclopaedia Britannica Educational Corp.
425 N. Michigan Ave.
Chicago, IL 60611
Black and white, 30 minutes, sound, 16mm, 1957.

The absorption of penetrating radiation by matter is described. Measurements of the absorption of beta and gamma rays, by solids, are made by means of Geiger-Mueller counter.

Engineering

The Breeder

Stuart Finley, Inc.
3428 Mansfield Road
Falls Church, VA 22041
Color, 23 minutes, sound, 16mm, 1978.

A comprehensive overview of the nuclear breeder and the many social, scientific, and environmental issues associated with nuclear power. Film recognizes the public apprehension and the unresolved problems. Present-day problems are discussed, including radioactive waste management, reactor safety, and the transportation of nuclear materials.

Nuclear Fusion

The Ultimate Energy

ERDA-TIC
P.O. Box 62
Oak Ridge, TN 37830.
Color, 28 minutes, sound, 16mm, 1976.

Discusses government-sponsored nuclear fusion research. Film begins with a question, What is nuclear fusion? The fusion process is described in nontechnical terms by leading physicists from a number of laboratories.

Nuclear Weapons

The Atomic Cafe

HBO Home Video
1370 Avenue of the Americas
New York, NY 10019
Color, 92 minutes, sound, Beta, VHS, 1982.

This is a humorous compilation of 1940s and 1950s newsreels and government films showing America's thoughts of the atomic bomb.

Nuclear Arms and Arms Control

United States Department of State
2201 C Street NW
Washington, DC 20242
Color, 25 minutes, sound, 1/2" VHS, 1984.

Students come to Washington to pose questions to a panel of specialists on nuclear arms and arms control. Issues discussed by the panelists include the difficulties in reaching an arms control agreement with the Soviet Union, the chances of a nuclear accident and of an intentional firing of a nuclear weapon, why the United States will not renounce the first use of nuclear weapons in the event of an attack in Europe, and if the U.S. development of nuclear weapons has made the Soviet Union more receptive to arms control.

Nuclear Strategy for Beginners

Time-Life Film and Video
100 Eisenhower Dr.
Paramus, NJ 07652
Color, 52 minutes, sound, 16mm, 1983.

Reviews four decades of the atomic age, with emphasis on the 50,000 nuclear weapons in the world. Stresses the debate as to whether nuclear weapons will begin or deter a nuclear war. Looks at the American and Soviet nuclear policy as well as the antinuclear movement.

Nuclear Warfare

The Atom and Eve

Green Mountain Post Films Inc.
P.O. Box 229
Turners Falls, MA 01376
Color, 15 minutes, sound, Beta, VHS, other formats available from the distributor, 1984.

An antinuclear war film.

Atomic Attack

International Historic Films
P.O. Box 29035
Chicago, IL 60629
Black and white, 50 minutes, sound, Beta, VHS, 3/4" U-matic cassette, 1950.

This is a drama on the bombing of New York City and the effects on a 50-mile area around the city.

Nuclear Countdown

Journal Films
930 Pitner Ave.
Evanston, IL 60202
Color, 28 minutes, sound, 16mm, 1978.

This film exposes the danger of the greatest threat facing mankind today. Emphasis is placed on the fact that lasting peace cannot be based on expansion of nuclear weapons. It traces the history of the nuclear arms race, documents international efforts of control and explains the dangers of both vertical and horizontal proliferation. It presents a basic theme of disarmanent.

Nuclear Defense at Sea

International Historic Films
P.O. Box 29035
Chicago, IL 60629
Color, 20 minutes, sound, Beta, VHS, 3/4″ U-matic cassette, 1987.

Studies the measures U.S. surface ships would take to survive a nuclear attack.

Nuclear Forces

National Educational Television, Inc.
WNET/Thirteen
Indiana University
Bloomington, IN 47401
Black and white, 20 minutes, sound, 16mm, 1966.

Examines the various means of delivering nuclear weapons to distant targets and protecting the delivery systems from surprise attack. Discusses how the vulnerability of bombers to nuclear attack has led to the development of protected missile sites. Reviews the defenses of the United States against attack.

Nuclear Nightmares: Wars That Must Never Happen

WNET/Thirteen Non-Broadcast
356 West 58th Street
New York, NY 10019
Color, 90 minutes, sound, Beta, VHS, 3/4″ U-matic cassette, 1980.

The film narrated by Peter Ustinov was taken at various military sites around the world. Shows changes of nuclear war and searches for factors that might lead to a nuclear war.

Nuclear War: A Guide to Armageddon

British Broadcasting Co.-TV
630 Fifth Ave.
New York, NY 10020

Color, 25 minutes, sound, 16mm, 1982.

Attempts to reveal the devastation of a one-megaton nuclear bomb exploding one mile above the center of a major city—London. By means of simulation, shows the effects of heat, blast and fallout, and assesses the effectiveness of measures to protect humans.

Nuclear Weapons—Can Man Survive?

King Features Entertainment
235 East 45th Street
New York, NY 10017
Color, 24 minutes, sound, Beta, VHS, 3/4″ U-matic cassette, 1983.

Film gives history of nuclear power since 1945 and examines the dangers and challenges of the nuclear age.

The Nuclear Winter: Changing Our Way of Thinking

Educational Film & Video Project
5332 College Avenue, Suite 101
Oakland, CA 94618
Color, 58 minutes, sound, 1/4″ and 1/2″ video, 1985.

A video production of a speech by Carl Sagan. Sagan explains how a nuclear winter could be initiated and the important aspects affecting the environment and man.

The Nuclear Winter: A Growing Global Concern

Educational Film & Video Project
5332 College Avenue, Suite 101
Oakland, CA 94618
Color, 20 minutes, sound, 1/2″ and 3/4″ video, 1985.

This is a compilation of film clips that examines how nuclear winter could occur, what it entails, its climatic consequences and short-term and long-term effects.

Radiation Carcinogenesis: The Hiroshima and Nagasaki Experience

University of Texas System Cancer Center
M.D. Anderson Hospital & Tumor Institute
Department of Medical Communications
Texas Medical Center
Houston, TX 77030
Color, 59 minutes, sound, 3/4″ U-matic cassette, 1981.

Discusses radiation-related malignancies resulting from the Hiroshima and Nagasaki bombings.

Radiation: Impact on Life

Bullfrog Films Inc.
Oley, PA 19547
Color, 23 minutes, sound, Beta, VHS, 3/4″ U-matic cassette, 1982.

Deals with the effects of high levels of radiation on the body as noted after the bombing of Hiroshima.

The War Game

Films, Inc.
1144 Wilmette Ave.
Wilmette, IL 60091
Black and white, 47 minutes, sound, 16mm, 1975.

Presents the consequences of nuclear war in dramatic and indelible fashion. Part 1 describes the precautionary evacuation plans and civil defense preparation, and Part 2 considers the possible responses to a nuclear attack and raises the question of retaliation. Other subjects include rehabilitating survivors—physically, mentally, and economically.

The World After Nuclear War—Nuclear Winter

University of Michigan
Michigan Media, 400 Fourth Street
Ann Arbor, MI 41803-4816
Color, 28 minutes, sound, 3/4″ U-matic, 1984.

Drs. Carl Sagan and Paul Ehrlich discuss with scientists how the use of a small fraction of the world's nuclear arsenal could trigger a nuclear winter and have devastating climatic effects. They stress that civilization as we know it could become extinct. The film stresses that the world's people need to recognize the potential catastrophe.

Nuclear Energy

The Atom and the Environment

Handel Film Corporation
8730 Sunset Boulevard
West Hollywood, CA 90069
Color, 22 minutes, sound, Beta, VHS, 3/4″ U-matic cassette, 1971.

Looks at the relationship between nuclear energy and ecology.

Learning About Nuclear Energy, 2d ed.

Encyclopaedia Britannica Educational Corp.
425 N. Michigan Ave.
Chicago, IL 60611

Color, 15 minutes, sound, 16mm, 1975.

Presents an overview of the production of nuclear energy. Society's responsibility for the safety of nuclear power is stressed.

The Nuclear Conspiracy

Vidmark Entertainment
2901 Ocean Park Boulevard
Santa Monica, CA 90403
Color, 115 minutes, sound, Beta, VHS, 1985.

Wife searches for her husband who disappears while investigating a nuclear waste shipment.

Nuclear Disintegration

Encyclopaedia Britannica Educational Corp.
425 N. Michigan Ave.
Chicago, IL 60611
Black and white, 30 minutes, sound, 16mm, 1957.

A film on original discoveries in the process of producing nuclear weapons. It includes Rutherford's discovery of indirect nuclear disintegration, Chadwick's discovery of the neutron, Einstein's mass energy equation, and the relation between different forms of nuclear energy.

Nuclear Energy

NETCHE (Nebraska ETV Council for Higher Education)
Box 8311
Lincoln, NE 68501
Color, 30 minutes, sound, 1/2" open reel (EIAJ), 3/4" U-matic cassette, 1976.

Shows how uranium and plutonium are fissioned, how the reaction is controlled, and how electrical energy is made from the energy produced.

Nuclear Energy

Encyclopaedia Britannica Educational Corp.
425 N. Michigan Ave.
Chicago, IL 60611
Black and white, 30 minutes, sound, 16mm, 1957.

Reveals how the fusion of light energy and the fission of heavy nuclei each liberates energy. Concept of positive and negative energy. Fission of uranium. How a chain reaction is obtained. Animated.

Nuclear Energy—Help or Hazard

University of Colorado
Educational Media Center
Boulder, CO 80309
Color, 29 minutes, sound, 16mm, 1979.

Presents the processes involved in the production of nuclear energy. Discusses uranium mining and milling. The nation's largest nuclear power plant, Trojan plant in Rainier, Oregon, is visited. There is also a short visit to the Doublet, Illinois, fusion reactor experiment conducted by General Atomic Corporation of San Diego, California.

Nuclear Energy: A Perspective

Modern Talking Picture Service
5000 Park Street North
St. Petersburg, FL 33709
Color, 28 minutes, sound, 3/4″ U-matic cassette, 1984.

Tells the story of uranium from the search for ore to fueling of nuclear reactors. Also looks at the future of breeder reactors.

Nuclear Energy: A Perspective

New York State Education Department
Center for Learning Technologies
Media Distribution Network
Room C-7, Concourse Level
Cultural Education Center
Albany, NY 12230
Color, 28 minutes, sound, Beta, VHS, 1/2″ open reel (EIAJ), 3/4″ U-matic cassette, 2″ quadraplex open reel, 1981.

Shows positive analysis of nuclear energy.

Nuclear Energy: The Question Before Us

National Geographic Society
17 and M Street, NW
Washington, DC 20036
Color, 25 minutes, sound, 16mm, 1981.

This film presents an objective view inside a nuclear energy plant. It shows how electricity is produced. The advantages and disadvantages of nuclear energy are shown, including the disposal of nuclear wastes. The role of government is stressed.

Nuclear Energy Training

NUS Training Corporation
910 Clopper Road
Gaithersburg, MD 20878-1399
Color, 60 minutes, sound, Beta, VHS, 1/2″ open reel (EIAJ), 3/4″ U-matic cassette, 1978.

Provides training system for nuclear license candidates and nuclear power plant technicians in 72 programs, grouped in seven modules. The seven modules—nuclear power orientation, basic nuclear power concepts, reactor operation, plant performance, radiation protection, water and waste treatment, instrumentation and operational analysis—are available individually.

Nuclear Power

Canadian Broadcasting Corp.
P.O. Box 500
Terminal A
Toronto, Ontario
Color, 28 minutes, sound, 16mm, 1980.

Discusses the pros and cons of replacing older methods of producing electricity with nuclear energy. Reviews the dangers of disposing of nuclear wastes and the means to ensure the security of neighborhood communities. The Canadian nuclear power system is reviewed.

Nuclear Power and You

Michigan Media
University of Michigan
400 Fourth Street
Ann Arbor, MI 48109
Color, 29 minutes, sound, 3/4" U-matic cassette, other formats available from the distributor, 1979.

Shows how nuclear energy works as a power source, its risks, and benefits involved.

Nuclear Power in the United States

Atomic Energy of Canada
Ottawa, Canada
Color, 28 minutes, sound, 16mm, 1971.

Describes the implementation of plutonium recycle programs and the thrust of the liquid metal fast-breeder. The total nuclear power industry is discussed based on advanced reactor concepts.

Nuclear Power Orientation

NSU Training Corporation
910 Clopper Road
Gaithersburg, MD 20878-1399
Color, 60 minutes, sound, Beta, VHS, 1/2" open reel (EIAJ), 3/4" U-matic cassette, 1978.

Multimedia training system for nuclear license candidates and power plant technicians.

Nuclear Power: Pro and Con

American Broadcasting Co. TV
1330 Avenue of the Americas
New York, NY 10019

Color, 50 minutes, sound, 16mm, 1979.

Explores both sides of the nuclear energy debate. Investigates proponents' claims of clean, cheap, and safe energy and the opponents' viewpoint of accidents, waste disposal and government subsidies that make nuclear plants appear less costly than they are.

The Nuclear Truth with Larry Bogart

Willow Mixed Media, Inc.
P.O. Box 194
Glenford, NY 12433
Color, 14 minutes, sound, Beta, VHS, 3/4" U-matic cassette, 2" quadraplex open reel, 1980.

Larry Bogart, antinuclear activist, offers a unique perspective on the history and future of the safe-energy movement.

Radiation and Your Environment

Educational Materials & Equipment Company
P.O. Box 17
Pelham, NY 10803
Color, 25 minutes, sound, Beta, VHS, 3/4" U-matic cassette, 1982.

Shows where radioactive emissions come from, their uses and dangers.

Radiation Redux

Media Business, Inc.
P.O. Box 718
Woodstock, NY 12498
Color, 6 minutes, sound, Beta, VHS, 1/2" open reel (EIAJ), 3/4" U-matic cassette, 1980.

Interweaves nuclear history, patriotic music, and words of nuclear activists with images of nuclear installations.

Radiation Safety Training for Support Personnel

NUS Training Corporation
910 Clopper Road
Gaithersburg, MD 20878-1399
Color, 23 minutes, sound, Beta, VHS, 1/2" open reel (EIAJ), 3/4" U-matic cassette, 1977.

Orientation in areas affecting support personnel, that is, recognizing warning signs, rules for working around radioactive materials, workers' obligations and rights, and proper accident response.

Nuclear Safety

Nuclear Safety Debate

Journal Films
930 Pitner Ave.
Evanston, IL 60202
Color, 26 minutes, sound, 3/4" U-matic cassette, 1/2" VHS, 1977.

Presents the debate between the industry that claims nuclear energy is the cleanest and safest form of energy and the ecologists who strongly disagree. Provides documents of the two viewpoints.

The Nuclear Watchdogs

Columbia Broadcasting System
383 Madison Avenue
New York, NY 10017
Color, 13 minutes, sound, 16mm, 1979.

A documentary about the safety of nuclear plants under construction. South Texas Nuclear Project is used to illustrate construction. Federal inspections to verify work by the South Texas Project were plagued with charges of faulty construction. Raises questions of the best ways to enforce federal inspections.

Radiation Safety—The Key To Contamination Control

University of Calgary
Communications Media
2500 University Drive NW
Calgary, AB
Canada T2N1N4
Color, 25 minutes, sound, Beta, VHS, 3/4" U-matic cassette, 1987.

Surveys the types of contamination and ways to prevent and control people's exposure to them.

Radioactive Contamination—Emergency Procedures for Transportation Accidents

Radiation Management Corporation
5301 Tacony Street
Box D5
Philadelphia, PA 19137
Color, 15 minutes, sound, Beta, VHS, 3/4" U-matic cassette, 1980.

Demonstrates the handling of victims, isolation of the contaminated area after determining radiation levels, and control of fire hazards after a highway accident.

Radioactive Contamination—Emergency Room Procedures

Radiation Management Corporation
5301 Tacony Street
Box D5
Philadelphia, PA 19137
Color, 20 minutes, sound, Beta, VHS, 3/4" U-matic cassette, 1980

Demonstrates the handling of an injured person who is contaminated. Shows how to protect hospital staff and ambulance personnel.

Nuclear Waste

Here Today—Here Tomorrow—Radioactive Waste in America

University of Minnesota
University Films and Video
1313 Fifth Street, SE
Minneapolis, MN 55414
Color, 30 minutes, sound, 1/2" VHS, 3/4", 1981.

This film looks at the current state of radioactive containment technology and plans for disposal. It gives a short history of the nuclear power industry and interviews researchers about the current state of storage technology and plans for the future.

Nuclear Waste Isolation: A Progress Report

Battelle Project Management Division
Office of Nuclear Waste Isolation, Communications Department
505 King Avenue
Columbus, OH 43210
Color, 25 minutes, sound, 16mm, 1980.

Discusses the plan proposed by the Office of Nuclear Waste Management for the disposal of radioactive nuclear waste. Shows the effects of heat and radiation on extremely stable geologic formations in various areas of the United States. Shows a possible site for disposal.

Accidents

Incident at Brown's Ferry

Time-Life Films
Time-Life Building
Rockefeller Center
New York, NY 10020

Color, 58 minutes, sound, 16mm, 1977.

The film attempts to provide answers to the question, How safe are nuclear power reactors? The nuclear fire at Brown's Ferry, Alabama, on March 22, 1975, is described. Animated sequences show how a reactor works and the engineering and licensing of nuclear plants. The film states what was done to improve safety at Brown's Ferry and cites other nuclear accidents.

Nuclear Fallout

Nuclear Fallout—Fiction and Fact

Hearst Metrotone News
235 E. 45th Street
New York, NY 10017
Black and white, 9 minutes, sound, 16mm, 1962.

Presents a report on the hazards of local and worldwide fallout. Explains the differences among three types of fallout. Shows that strontium 90 and cesium 137, the two most dangerous elements, have not produced a significant risk up to 1962.

Radioactive Fallout and Shelter

Ohio Department of Health
Columbus, OH 43215
Color, 28 minutes, sound, 16mm, 1960.

Discusses the effect of radiation on people and emphasizes the need to protect against it.

Nuclear Radiation Emergency

Media Project Inc.
P.O. Box 4093
Portland, OR 97208
Color, 14 minutes, sound, Beta, VHS, 1/2″ open reel (EIAJ), 3/4″ U-matic cassette, 1980.

A training program for nurses and medical staff on emergency room procedures in handling and treating victims in case of a nuclear radiation accident.

Radiation Effects

Telstar Productions Inc.
366 North Prior Avenue
St. Paul, MN 55104

Black and white, 30 minutes, sound, Beta, VHS, 1/2″ open reel (EIAJ), 3/4″ U-matic cassette, 1967.

Illustrates immediate and long-range effects of exposure to radiation.

Radiation Safety

Indiana University
Audio-Visual Center
Bloomington, IN 47405
Color, 13 minutes, sound, 3/4″ U-matic cassette, 1981.

Used for training persons who will use radioactive materials in research.

Radiation Therapy

Trainex Corporation
P.O. Box 116
Garden Grove, CA 92642
Color, 25 minutes, sound, Beta, VHS, 3/4″ U-matic cassette, 1977.

Discusses the kinds of radiotherapy used in the treatment of cancer.

Political

Lovejoy's Nuclear War

Green Mountain Post Films
P.O. Box 177
Montague, MA 01351
Color, 60 minutes, sound, 16mm, 1975.

This film reports an incident in the protest movement against the construction of nuclear power plants. Sam Lovejoy toppled a steel weather tower erected by Northeast Utility Company in Massachusetts as part of its construction of a nuclear power plant. Film consists of interviews, news clippings, and part of speech by President Ford. The film presents a provocative viewpoint of social protest and civil disobedience.

No Act of God

Bullfrog Films
Oley, PA 19547
Color, 28 minutes, sound, 16mm, 1978.

Presents the danger to society from the use of nuclear energy. Film suggests limiting its use. Testimony by the noted Swedish Nobel Laureate Hannes Alfven and others presents the problems of using nuclear energy, but solutions are not given.

Nuclear Power in World Politics

National Broadcasting Company
30 Rockefeller Plaza
New York, NY 10020
Color/black and white, 20 minutes, sound, 16mm, 1966.

Shows how nuclear weapons are a primary force in world diplomacy. Reveals the destructive capacity of nuclear weapons. Chet Huntley interviews key men in many countries regarding crucial decisions and danger of nuclear buildup.

Nuclear Power: The Vital Debate. Part 1, How Safe Are America's Atomic Reactors. Part 2, A Small Core of Blackmail

Impact Films
144 Bleecker St.
New York, NY 10012
Part 1: Color, 28 minutes, sound, 16mm, 1972.
Part 2: Color, 27 minutes, sound, 16mm, 1972.

This British film presents an antinuclear viewpoint in the nuclear power debate. Part 1 centers on criticism of the emergency core cooling system and the conduct of the former Atomic Energy Commission. Part 2 centers on the possible theft of plutonium by terrorist criminals. Particular concern is given to the way plutonium is transported. Stresses the international aspects of the problem.

Nuclear Reaction in Wyhl

Green Mountain Port Films
P.O. Box 229
Turner Falls, MA 01376
Color, 15 minutes, sound, 1/2″ Beta and 3/4″ U-matic cassette, 1976.

A film chronicling the spontaneous European sit-ins at nuclear power sites in 1975. Uses actual 8mm footage filmed by the protesters.

Glossary

alpha A positively charged particle given off by radioactive substances. It consists of two protons and two neutrons. It is less penetrating than a beta ray.

atom The smallest particle into which an element can be chemically divided.

background radiation The radiation in the natural human environment originating from cosmic rays and from the naturally radioactive elements of the earth.

Becquerel rays Invisible rays emitted by radioactive substances such as uranium, radium, and thorium.

beta rays Rays given off by radioactive substances consisting of electrons that move with high velocities up to 180,000 miles per second.

chain reaction A self-sustaining reaction occurring in nuclear fission when the number of neutrons released equals or exceeds the number of neutrons absorbed plus the neutrons that escape from the reactor.

collective dose The sum of the individual doses received by each member of a certain group of a population. It is calculated by multiplying the average dose per person by the number of persons within a specific geographic area.

core The central part of a nuclear reactor that contains the fuel and produces the heat.

cosmic rays Rays of extremely short wavelength and great penetrating power that bombard the earth from outer space.

curie (Ci) The unit used in measuring radioactivity. The quantity of any radioactive atom of a specific nuclear constitution in which the number of disintegrations per second is $3{,}700 \times 10^{10}$, approximately that occuring in one gram of radium.

decay heat Heat produced by the decay of radioactive particles.

deuterium Hydrogen isotope having an atomic weight of 2. It is called heavy hydrogen and with oxygen forms heavy water.

digital radiography The conversion of X-rays through electronic impulses in a picture on a television screen.

dose The energy imparted to matter by ionizing radiation per unit mass of irradiated material at a specific location. The unit of the absorbed dose is the rad.

dyne A unit of force. One dyne equals the force necessary to give 1 gram mass an acceleration of 1 centimeter/(second) (second).

electron An elementary particle that is the negatively charged constituent of ordinary matter. It is the lightest known particle that possesses an electric charge.

enriched uranium Uranium in which the percentage of the fissionable isotope uranium 235 has been increased above the 0.7 percent contained in natural uranium.

erg A unit of energy. One erg equals the work done when a force of 1 dyne is applied through a distance of 1 centimeter.

fission The splitting of a heavy nucleus into two roughly equal parts, accompanied by the release of a relatively large amount of energy and frequently one or more neutrons.

fuel damage The failure of fuel rods and the release of the radioactive fission products trapped inside them. Fuel damage can occur without a melting of the reactor's uranium.

fuel rod A tube containing fuel for a nuclear reactor.

gamma rays High-energy, short-wavelength electromagnetic radiation emitted by a nucleus. Gamma radiation usually accompanies alpha and beta emissions.

gigawatt A million watts.

gigawatt-hour One million kilowatt-hours.

gram A unit of mass. One gram equals 1/1,000 kilogram.

half-life The time in which half the atoms of a given quantity of a particular radioactive substance disintegrate to another nuclear form.

health physics The practice of protecting humans and their environment from the possible hazards of radiation.

heavy water (D_2O) Water produced when oxygen is combined with deuterium, an isotope of hydrogen that contains one extra neutron.

ionizing radiation radiation capable of displacing electrons from atoms. The process produces electrically charged atoms such as X-rays, alpha, beta, and gamma rays.

irradiation Radiation used in medicine to treat diseased tissue by exposure to X-rays, ultraviolet rays, radium, or some other form of radiant energy.

isotope One or two or more forms of an element that differ in atomic weight. Nuclides have the same atomic number and the same number of protons, but a different number of neutrons in the nucleus.

isotopic Of or relating to isotopes.

light water Normal water (H_2O), as distinguished from heavy water (D_2O).

light water reactor A reactor in which ordinary water (H_2O) is used as a coolant and moderator.

long-lived nuclides Radioactive isotopes with half-lives greater than about 30 years. Most long-lived nuclides of interest to waste management have half-lives longer than thousands of years.

low-level waste Waste containing types and concentrations of radioactivity such that minimum shielding is required.

meltdown The melting of fuel in a modern reactor after the loss of coolant water.

mesotron An unstable particle larger than an electron that is formed by the disintegration of cosmic rays.

micron A particle having a diameter between .01 and .0001 millimeter.

neutron An uncharged elementary particle of an atom that has approximately the same mass as a proton. Neutrons sustain the fission chain reaction in nuclear reactors.

noble gases Gases such as helium, neon, and argon that are inert chemically.

nuclear fuel cycle The series of steps from the mining of uranium to the final stage in supplying fuel for the nuclear power generator and the final disposal of radioactive waste.

nuclear fuel reprocessing The processing of irradiated (spent) nuclear reactor fuel to recover useful materials as separate products, usually involving separation into plutonium, uranium, and fission products.

nuclear reactor A device in which a fission chain reaction is initiated, maintained, and controlled.

nuclide A nuclide is an individual atom of given atomic weight and energy content that exists for a measurable length of time and has a distinct nuclear structure.

palliation The use of radiation to treat cancer at an advanced stage in order to prolong the life of the patient.

picocurie (pCi) A picocurie is one-trillionth of a curie.

plutonium A radioactive element with an atomic number of 94. Its most important isotope is fissionable plutonium 239, produced by neutron irradiation of uranium 238.

primordial radionuclide A radionuclide that exists from the time of the origin of the world. Its half-life extends into billions of years.

proton A fundamental part of the nuclei of all atoms that carries a unit positive charge of electricity.

rad (radiation absorbed dose) A rad of radiation exposure is equal to 100 ergs of radiation energy deposited in one gram of matter.

radiation The term used to describe the process in which energy, in the form of rays of light or heat, is sent out from atoms and molecules as they undergo change.

radioactive decay Property of undergoing spontaneous nuclear transformation in which nuclear particles or electromagnetic energy are emitted.

radioactivity The disintegration of unstable atomic nuclei, accompanied by the emission of radiation.

radioisotope An unstable radioactive isotope of an element that decays or disintegrates spontaneously, emitting radiation. More than 1,300 natural and man-made isotopes have been identified.

radionuclide An unstable nuclide of an element that decays or disintegrates spontaneously, emitting radiation.

radon A source of natural radiation, radon is a tasteless, odorless, invisible gas seven times heavier than air.

recycled fuel Uranium and plutonium recovered for reuse as reactor fuel elements.

rem (roentgen equivalent man) A dosage of any ionizing radiation that will produce a biological effect approximately equal to that produced by one roentgen of X-ray or gamma ray radiation. The rem is defined as a dose of a particular type of radiation required to produce the same biological effect as one roentgen of gamma radiation.

roentgen (R) A measure of the ability of gamma or X-rays to produce ionization in air. One roentgen corresponds to the absorption of about 86 ergs of energy from X-ray or gamma-ray radiation per gram of air. The corresponding absorption of energy in tissue may be from one-half to two times as great.

short-lived nuclides Radioactive isotopes with half-lives not greater than 30 years, such as cesium 137 and strontium 90.

sievert (Sv) The sievert is a unit for measuring the effectiveness of various ionizing radiations in causing harm to tissue.

site emergency Declared by the utility when an incident at a nuclear power plant threatens the uncontrolled release of radioactivity into the immediate area.

somatic cells Any cell of an organism that becomes differentiated into the tissue of the body.

tailing The waste or refuse from various processes of milling ores.

uranium A natural radioactive element with the atomic number 92 and an atomic weight of about 238. The two principal naturally occurring isotopes are uranium 235 (fissionable) and uranium 238.

uranium oxide UO_2 A chemical compound containing uranium and oxygen that is used as a fuel in nuclear reactors.

weighted dose The weighted dose is known as the dose equivalent and is measured in units called sieverts. The weighted dose is required because each of the nuclear particles has a different effect. For example, alpha particles are much more damaging than beta or gamma radiation, so the dose needs to be weighted for its potential to do harm. Alpha radiation is given 20 times the weight of the others.

whole-body dose A radiation dose to the whole body.

Index